P9-DHD-556

THE EVOLUTIONARY WORLD

Also by Geerat Vermeij

Nature: An Economic History

Privileged Hands: A Scientific Life

A Natural History of Shells

Evolution and Escalation: An Ecological History of Life

Biogeography and Adaptation: Patterns of Marine Life

THE EVOLUTIONARY WORLD

How Adaptation Explains Everything from Seashells to Civilization

GEERAT J. VERMEIJ

THOMAS DUNNE BOOKS
ST. MARTIN'S PRESS NEW YORK

THOMAS DUNNE BOOKS.
An imprint of St. Martin's Press.

www.thomasdunnebooks.com
www.stmartins.com

ISBN 978-0-312-59108-3

First Edition: December 2010

10 9 8 7 6 5 4 3 2 1

To Alfred G. Fischer and Egbert G. Leigh Jr.,
two great mentors

CONTENTS

ACKNOWLEDGMENTS

Writing is often thought of as a lonely pursuit, an activity best practiced in seclusion, accompanied only by books and, in my case, by classical music. But there is much more to constructing a readable narrative than mere writing. It takes a community of colleagues, friends, and institutions to shape ideas, sharpen arguments, acquire the necessary facts, create a climate suitable for reflection, provide security of employment and daily life, and see a manuscript through to publication.

Dozens of people have influenced my thinking and discussed ideas with me in highly constructive ways. Foremost among them is my wife Edith, whose wide knowledge, copious curiosity, intelligent analysis, probing questions, and great companionship have immeasurably enriched my intellectual and personal life. Egbert Leigh, an annual visitor to our house for decades, helped me appreciate the importance of interdependence in the evolutionary affairs of living things, and expanded my horizons in fields ranging from genetics to ecology and ethics. Many of my students have become enthusiastic, stimulating colleagues and contributors, including Mark Bertness, Kevin Carpenter, Melissa Frey, Gregory Herbert, Michael Kirby,

Halard Lescinsky, and Peter Roopnarine. Among the many other colleagues who have benefited me with their insights, expertise, and criticism are James H. Brown, Richard Cowen, Cliff Cunningham, Gregory Dietl, Robert Dudley, Frietson Galis, Rick Grosberg, Jeremy Jackson, Christine Janis, Conrad Labandeira, David Lindberg, Ryosuke Motani, Tatsuo Oji, Jeanine Olsen, Kevin Peterson, Theunis Piersma, Nick Pyenson, Rafe Sagarin, Menno Schilthuizen, Lars Schmitz, Paul Selden, Richard Strathmann, Peter Wainwright, and Frank Wesselingh.

Over the past fifty years, I have accumulated a vast Braille library, which consists almost entirely of extensive notes I have transcribed from the thousands of books and scholarly articles that others—first my mother and my brother Arie, then Bettina Dudley at the University of Maryland, my wife Edith, and Janice Cooper at the University of California, Davis—have read aloud to me. I am hugely indebted to these readers, who have not only had to wrestle with technical terms, tables of numbers and names, bad writing, and mathematical expressions, but also with many languages in addition to English. Edith and Janice have also helped with my correspondence, and Edith has worked with me at field sites and museums around the world.

Museums have provided invaluable access to their collections, libraries, and curators. A Temminck Fellowship allowed me to spend some wonderful weeks of research on fossils at Naturalis (National Museum voor Natuurlijke Historie) in Leiden. I am also grateful to the California Academy of Sciences in San Francisco, where I have done extensive research and where, as a science trustee, I have learned much about education in a museum setting.

The University of California, Davis, has given me the luxury of freedom and time to develop my ideas, expose those ideas to my gifted colleagues and to the students I teach, and conduct research on topics of my choosing. I am particularly grateful to my department chairs and other officials for sparing me from most administrative duties, to which I am singularly unsuited.

A fellowship from the John D. and Catherine T. MacArthur Foundation not only gave me financial support but also the intellectual encouragement to explore ideas and research directions that more risk-averse sources were unlikely to fund. I am also indebted to the National Science Foundation for supporting some of my more conventional research.

Small conferences have proven to be ideal settings for exchanging ideas and trying out new ones. I am grateful to have participated in such gatherings, including the Geological Society of America's Penrose conference on tectonics in St. George, Utah, in 2005; meetings of CORONA (The Committee on Research on the North Atlantic) from 2002 to 2006, held at Reykjavik (Iceland), Roscoff (France), and Tavira (Portugal), among other places; the celebration of Carolus Linnaeus's three hundredth birthday anniversary, held in Amsterdam in 2007, organized by the Linnean Society and the Royal Netherlands Academy of Sciences; and the celebration of Charles Darwin's two hundredth birthday anniversary, held in Stockholm in 2009, sponsored by the Swedish Academy of Sciences and the Swedish Museum of Natural History. Other notable events were a celebration of Adolf Seilacher's eightieth birthday, a symposium on form and function of fossil organisms held at Yale University in New Haven in 2005; a conference on metahistory in 2008 in Honolulu, sponsored by the Santa Fe Institute; and the Paleontological Society's Short Course on Conservation Paleobiology held in 2009 in Portland, Oregon.

I am enormously fortunate to have first-rate editorial support. Janice Cooper has seen to it that my written work goes out with as few errors as possible. Peter Tallack, my agent, not only knows the world of science publishing intimately, but has given me encouragement and superb advice about crafting a credible book proposal and manuscript. Peter Joseph, my editor at Thomas Dunne/St. Martin's, has a remarkable sense of how to make abstract ideas come alive, which questions a reader might ask, which details to fill in or leave out, and when to call for clarification. He is, quite simply, the best editor I have ever had.

PREFACE

This book is a personal, and in many ways an unconventional, look at evolution, an idea so powerful, so beautiful, and so far-reaching in its implications for human existence, that every educated person can be enriched and enlightened by it. Evolution—descent with modification—is a concept that organizes, explains, and predicts a multitude of unconnected facts and phenomena of life in nature past and present. It provides a coherent framework for understanding where we came from, where we are going, and how we and the rest of living nature create bewildering complexity, a world of meaning, and surpassing beauty. Quite simply, evolution has outgrown its original home in biology and geology. It is the foundation of a worldview in which environments, genes, organic architecture, physiology, chance, the economic struggle for life, and historical narrative come together to illuminate how we live in the world.

Yet evolution is often misunderstood, misconstrued, despised, and even denied by hundreds of millions of people. Skeptics harbor legitimate questions and reservations as well as ill-founded grievances. Some find random mutation and selection insufficient to account for adaptation. Others underestimate, or seem unaware of, the power of living and inanimate parts of nature acting together to create new structures with novel properties, and therefore wonder whether evolution can explain complexity, especially the origin of the mental and moral traits that seem to set humans apart from other forms of life. Many reject a world in which order, meaning, and beauty arise unintentionally through the action of simple, observable processes operating over eons of time. For them, evolution taking place without the initiation or intervention of a supernatural being robs life of all purpose and meaning and rips all the moral and ethical fabric from human society. To these skeptics, science in general, and evolutionary science in particular, is cold, clinically sterile, impersonal, and emotionally impenetrable.

The challenge for scientists like me, and one of the goals of this book, is not only to demystify evolution, but also to show how understanding its mechanisms and consequences yields an emotionally satisfying, aesthetically pleasing, and deeply meaningful worldview in which the human condition is bathed in a new light. In his book *Darwin's Dangerous Idea,* the American philosopher Daniel Dennett characterized evolution as "universal acid" to emphasize the power of evolutionary thinking to penetrate every nook of human knowledge. But this is a grim image, a metaphor that calls to mind the satanic power feared by doubters and deniers. Evolution is not some corrosive agent, but a universal elixir that enriches those willing to taste it.

I want readers of this book to come away with a firmer grasp of the grandeur of evolution—its facts, mechanisms, puzzles, directions, and implications—but above all, I want them to glimpse the love of the living world that an exploration of evolutionary concepts can elicit.

There may not be a gene for the appreciation of nature, but my family's enjoyment and knowledge of things in the wild were certainly infectious. Growing up in the Netherlands, I was drawn to the meadows and alder thickets of the low-lying countryside around Gouda, to the sound of the wind through the poplars along the dike, and to the prospect of an approaching thunderstorm. I spent many happy hours sorting the shells I gathered on the wide sandy beach at Scheveningen. On those all-too-rare weekends when I was allowed to come home from boarding school, there would often be a new book about nature, laboriously hand transcribed into Braille by my mother, waiting for me.

After immigrating to the United States in 1955, I began to collect seashells in earnest. At first, shells appealed to me chiefly as exquisite hand-sized pieces of sculpture, embodying the perfection and harmony I expected to find in the nonhuman world of nature; but as curiosity grew into passion, I began to realize that some places, notably the warm seas of the tropics, produced wonders of such beauty and intricacy that I could scarcely imagine how living clams and snails could fashion them, especially given the much less ornate shells familiar to me from the North Sea. As I contemplated my shells and the other objects I was collecting—feathers, seeds, dried plants, minerals, and even samples of wood—the ideas about evolution that I encountered in books about Charles Darwin and other great naturalists of the past seeped almost unnoticed into a receptive mind that was hungry for big ideas.

When I arrived at Princeton as a freshman in 1965, brilliant and generous professors introduced me to evolution as science. Through classroom lectures, after-class discussions, and above all field trips—to New Jersey's shore and pine barrens, New Hampshire's Mount Washington, the Paleozoic rocks of Pennsylvania and New York, and the forests and mountains of Costa Rica—they impressed upon me the reality that living things are engaged in what Darwin called the struggle for life, which is the centerpiece of his theory of evolution by natural selection. Shells were no longer mere variations on a theme of spiral architecture; they told stories of lives led in places near and far and at times often remote from our own. Their shapes reflected evolutionary heritage as well as the challenges and opportunities to which shell builders were adapted. By piecing their stories together, it would be possible to reconstruct a history of evolving life and its ever-changing surroundings. I found this to be a thrilling prospect: I could combine a passion for nature, including my particular infatuation for all things molluscan, with a search for big explanatory ideas in science.

I was hardly the first to follow the path from a childhood love of nature to the more disciplined endeavor of evolutionary science. Many of the great evolutionists, from Charles Darwin to Harvard biologist and ant specialist Edward O. Wilson, built their illustrious discoveries on a foundation of observing and collecting the productions of nature. But for them, observation largely meant seeing. My sensory world, by contrast, was one of touch, sound, odor, and taste. What little vision I had at birth disappeared by design during the last of many operations on my glaucoma-affected eyes at the academic hospital in Utrecht, three months shy of my fourth birthday. Blindness came early enough that the parts of my brain where visual signals are integrated into images could be retooled to interpret the signals from my remaining senses. Except for vivid memories of colors, all my encounters with nature and my entire education proceeded in the absence of light perception.

My scientific outlook has been largely shaped by exposure to a great variety of environments—forests, reefs, mudflats, swamps, deserts, tundra, wave-swept rocky seashores, high mountains, fossil sites, remote islands, and natural-history museums—and to a vast literature that led me on long intellectual excursions into history, economics, and every discipline in science. Always I am driven by insatiable curiosity, by the overwhelming desire to make sense of the astounding diversity that evolution has wrought and that I find irresistible.

The phenomenon at the core of my evolutionary worldview is

adaptation, the good fit between organisms and their surroundings. To some readers, the centrality of adaptation in evolution will seem obvious; to others it seems wrong. I believe it is neither. Adaptation is the universal condition of living things, but it raises profound questions that deserve careful scientific scrutiny. How is adaptation achieved? Which features contribute to a well-functioning body or group? How are adapted states transmitted from the ancestral state to adapted descendants? And how might the perpetual struggle for life lead to a predictable trajectory of history? We cannot automatically assume that everything has a meaning or a function, but neither can we dismiss the good fit between organism and environment as an incidental and minor outcome of evolution. It is my aim in this book to explain how adaptation accounts for the major features of evolution and for life's history, including our own.

A familiar example illustrates two evolutionary phenomena—emergence, or synergy, and feedback—that are fundamental to adaptation and central to this book. An oak tree is an integrated living structure well suited to live on land and to carry out photosynthesis, the production of living tissues from water, nutrients, carbon dioxide, and sunlight. The roots, acting in partnership with specialized fungi, mine nutrients and water from the soil. The tree's trunk resists the force of gravity and lifts the leaves high enough to intercept sufficient light. The mature leaves are chemically defended by tannins against herbivores, but that defense also makes them inviting targets for gall wasps, which can tolerate those tannins. The seeds—acorns—support a cast of consumers, especially jays and squirrels, which bury some of the crop for later use. The cached acorns that remain uneaten grow up to become the next generation of oaks.

All the parts of the oak—the roots, trunk, leaves, and acorns—are well adapted; but in isolation, neither they nor their functions can be fully understood. The tree is more than the sum of its parts; it is an emergent whole, the integrated product of parts working together.

The oak is also part of a larger emergent community of pests and helpers, whose adaptations all depend on one another and on those of the oak. The parts of the tree interact with each other economically; so do all the species competing and cooperating with the oak. Consumers affect producers—oaks, in this case—and producers affect consumers through feedback; causes and consequences intertwine. Adaptation is, in other words, an economic phenomenon, which cannot be grasped without taking into account the ways in which individual organisms, and the parts of organisms, vie for resources and work together to create an emergent, adapted whole.

Three-and-a-half billion years of evolution have yielded a richness of adaptive solutions to problems ranging from the everyday threats of weather or being eaten to the extraordinary events of mass extinction. It is therefore not surprising that the processes and history of adaptation explain the origins and success of our characteristics, including our social values. So many adaptive solutions have been tested by living things under so many conditions, most of which are created by organisms themselves, that we can turn to this wealth of evolutionary "experience" to inform our own actions and decisions. In fact, it is the unity of the adaptive worldview, the emergence of so many unexpected parallels between the workings of nature and the affairs of human societies, that I find deeply satisfying and that motivates this book.

The economic and historical perspectives included in this book have led me to consider questions that few evolutionary biologists have confronted. I ask how adaptation by organisms parallels the scientific way of knowing, why no adaptation or scientific hypothesis can ever be perfect, how living things and human societies can incorporate very rare or even unprecedented challenges into an adapted structure, how radically new adaptive solutions arise, how short-term harm can turn into long-term opportunity through adaptation and feedback, and how patterns of history emerge despite the pervasive presence of chance and randomness. Along the way, I explore similarities between genomes and languages, the contrasting natural economies of islands and continents, the emergence and importance of human values, the long-range consequences of global warming, and the perils of monopoly. There are implications for education, our system of laws, and economic growth.

Although the history of ideas can be an effective means of exploring a subject like evolution, I have chosen instead to lay out the core principles of evolution and adaptation as we understand these principles today. In science, as in other disciplines, the route to current understanding is often circuitous, with many detours, deep ruts, and wrong turns. It is often necessary to hack through thickets of controversy and to clear away stands of stale semantics. These activities are best left to academic journals, where it is indeed important to thrash out arguments for distinguishing sound inference from false assumptions, misleading statements, and wishful thinking. Although I shall often refer to alternative viewpoints throughout this book, my chief aim is to present a coherent yet expansive account of evolution and its implications. I am a historian of life, not of science.

When evolution and adaptation are conceived of in utilitarian terms, some readers may feel that the magic of life is lost and that the worst fears of evolution as a heartless process devoid of all higher and meaning and purpose will come true. I maintain the contrary, however—that understanding where life comes from and how it has adapted adds value to the love and admiration we feel for our fellow creatures. Rather than detracting from the central role of meaning and purpose in our lives, evolutionary theory and history provide a solid foundation for it. Adaptive evolution explains and illuminates what it means to be alive.

THE EVOLUTIONARY WORLD

CHAPTER ONE

The Evolutionary Way of Knowing

Squirrel Island is one of those iconic places on the coast of Maine where an idyllic landscape of meadows, mossy slopes, and fragrant forests of spruce and pine meets the sea. A short boat ride from Boothbay Harbor, it was the perfect spot to spend time with my girlfriend Edith as we picked berries, listened to warblers singing high in the canopy, and examined all that grew in this tame fragment of northern nature.

But it was the seashore that brought us here. A luxuriant cover of slippery rockweeds, their air-filled bladders crackling under the pressure of our boots, gave way to mats of blue mussels and then to stands of waving kelp fronds as we descended to the low water mark. All the expected inhabitants of the rocky shore were here: spiny green sea urchins nestled in pools and among the kelp holdfasts, large hairy horse mussels crammed into sand pockets around boulders, rock crabs and green crabs hiding beneath sponge-lined ledges, sluggish dog whelks massing in crevices or perched atop prey barnacles, limpets clinging to and scraping crusts beneath rocks, and three kinds of periwinkle feasting on seaweeds. Gulls, always on the lookout for crabs, patrolled overhead. To judge from their large size and the many closely spaced growth interruptions on their weather-beaten shells, the blue mussels had been living their sedentary lives on this shore for decades. To us as naturalists, the shore and its surroundings exemplified the serenity and beauty that give nature and life meaning. To us as scientists, they posed questions and invited inquiry into the ways of the living world.

Grinding winter ice, breaking waves, intense competition for light and space, and a variety of predators—crabs, lobsters, sea stars, whelks, fish, and gulls—are part of life for every creature on this shore. Yet despite these dangers and despite the conflicting interests of the seaweeds and animals in this community, the species are well adapted to each other and to their rigorous physical surroundings. The secure attachment by flexible threads secreted from the foot protects mussels from being dislodged by shifting ice or powerful predators from rocks. Even when the threads fail and the mussels become detached, the bivalves can produce new threads and refasten themselves if they land in the right place. A mussel's two shell valves can shut tightly around the soft tissues, so that would-be attackers homing in on the chemical plume that a mussel releases while actively feeding and respiring loses the scent and finds another victim. When snails like dog whelks and periwinkles become dislodged, they quickly withdraw the delicate soft parts—head, foot, and other organs—far back in the shell, whose opening is sealed with a flexible yet tightly fitting doorlike device. Predatory sea stars, when directly contacting potential prey snails, often elicit a vigorous escape response. Sponges beneath overhangs secrete chemicals that enemy sea slugs find repellent. I could fill a book documenting the adaptations of every species on this relatively simple shore, let alone every living and fossil species on the planet.

The fact of adaptation—the good fit between organism and environment—was already well established by the time Charles Darwin and his collaborator and competitor Alfred Russel Wallace ushered in the evolutionary age with the presentation of their paper before the Linnean Society in London in 1858. All living things, including humans, effectively perform the essential functions of life—growth, metabolism, food intake, defense, maintenance of the body, and reproduction—in the places these organisms inhabit. Darwin and Wallace proposed one mechanism for adaptation, based on an idea borrowed from the English political economist Thomas Malthus—the number of offspring produced in a population exceeds the number surviving to reproduce as adults. Survivors possess heritable characteristics—those passed from generation to generation—that enable living individuals to meet the challenges and capitalize on opportunities in their environment more effectively than the traits of those individuals that die before reaching maturity. Adaptation thus involves a selective process, which was called natural selection by Darwin, resulting from the culling of inadequately performing individuals in the struggle for life.

Bipedal locomotion—running and walking on two legs—is a good example of human adaptation. The bipedal condition evolved from the ancestral four-footed condition many times, as in kangaroos and several lines of dinosaurs, but in primates it is restricted to our own species *(Homo sapiens)* and other species of *Homo,* dating back to perhaps as early as two-and-a-half million years ago. Among running animals in general and bipeds in particular, humans are unusual in that they can run for long distances, up to six miles (ten kilometers) per day for someone in good physical health. Only a few other mammals—wolves, hyenas, and African dogs that hunt in packs, as well as migrating horses and African wildebeest—engage in comparable endurance running, but these animals run on all fours. Our two-legged, long-distance running is made possible by modifications in the leg bones, joints, tendons, muscles, and even our skin. Unlike other running mammals, we achieve high speed with a long stride of up to three-and-a-half meters (more than ten feet) in highly trained athletes. Traits contributing to this long stride include unusually long hind limbs, a relatively short and lightweight foot, the unique presence of the Achilles tendon connecting the heel with muscles in the foot, and short bundles of muscles that produce the forces necessary for sustained running. All the joints in our legs, from the foot to the pelvis, are conspicuously enlarged compared to those of apes and compared to the joints in our arms. This enlargement enables the body to absorb the strong forces produced when the feet hit the ground as we run. The copious heat generated by our muscles during running is lost through the skin, which in humans has unusually abundant sweat glands through which evaporation takes place. Human skin is also peculiar in lacking insulating hair over most of its surface. These and other departures from the primate norm represent adaptations that enabled our immediate ancestors to compete effectively with other predators for scarce, protein-rich meat in hot open country. Other distinctive human adaptations—a large brain, delayed sexual maturity, the use of long-distance weapons, and the domestication of animals and plants—came much later. In the early history of our genus *Homo,* however, running-related modifications gave us a decisive competitive edge in a place where the struggle for life was particularly intense.

Not every aspect of an organism represents an adaptation. Many features are simply expressions of how an individual grows, or how it is put together, much as the seam of a plastic cup indicates how the cup was manufactured. Most shells, for example, grow in the shape of a spiral, reflecting a pattern of growth in which new shell material is added only at the expanding end of what is effectively a conical enclosure. The spiral can be variously modified in adaptive ways, ranging from the loose, rapidly expanding spiral of a clam shell or a cap-shaped limpet to the tightly coiled spiral of many snails; but the spiral form itself is not a direct expression of adaptedness. Likewise, the growth marks on the exterior of a mussel shell or in the bones of reptiles and some dinosaurs indicate periodic interruptions in growth rather than some subtle adaptation, but here again many animals have fashioned growth marks into larger features that impart greater shell strength and other protective functions.

Individual traits may even be harmful in some situations. An actively feeding mussel, which pumps water through its gills in order to extract food particles as well as oxygen, necessarily releases metabolic products with the expelled water. These are beacons for potential predators like sea stars and snails. The disadvantages associated with this inadvertent advertisement must be compensated for by an ability to recognize the nearby presence of hungry predators and to respond quickly by shutting the shell temporarily so that the scent disappears.

The catalog of potentially harmful traits in humans is also long. In the human eye, for example, there is no vision in the hole through which the optic nerve passes from the light-gathering surface of the retina to the brain. There is thus a blind spot located some thirty degrees to the right of the point of focus of the right eye, and another thirty degrees to the left of the focal point of the left eye.

It may even be that most evolutionary changes are neutral or nearly neutral, and therefore impervious to natural selection. In all organisms, genetic material (DNA and RNA) consists of helically wound chains of nucleotides of four kinds: adenine, cytosine, guanine, and thymine (or uracil). Some of this genetic material provides instructions for building proteins, which consist of long chains of twenty kinds of amino acids. In the sequence of three nucleotides that specifies a given amino acid, the third nucleotide of the sequence is typically silent—that is, a switch from one nucleotide to another in this position leaves the amino-acid specification unchanged. These alterations are likely to be common, but they have no effect on an organism's adaptation; they are thus effectively neutral. Even mutations with mildly deleterious effects may become established under some conditions. In very small populations of five or fewer individuals, a mutation with a tiny disadvantage can persist because a random fluctuation in population size by one or two individuals has a much larger effect on the fate of that mutation than selection does on the trait that the mutation specifies. If species often originate as tiny populations of this size, as many biologists believe, this so-called random fixation through genetic drift may be an important non-adaptive cause of genetic change. In still other cases, a new mutation or a new trait may be protected from selection because it is expressed in a very safe environment, as in the embryo inside the mother, where it is sheltered from the usual agencies comprising the struggle for life. In such situations, traits are relatively free to vary, and they therefore provide the raw material on which the agencies responsible for natural selection can act.

The features of an organism may be nothing more than by-products of construction, or they may be functionless legacies of the past, but living things are adapted units whose parts and characteristics enable them to thrive and propagate. Generations of such successful organisms form a lineage, whose members have accumulated enough adaptive information about their surroundings to pass on their traits and the instructions for those traits to the next generation. A lineage can persist for millions of years as long as its members keep working adequately. Often, a lineage divides, giving rise to two or more lineages in an evolutionary branching event. All the lineages of life—ancient as well as modern—can thus be thought of as forming an evolutionary tree, whose innumerable twigs can be traced back through time to a single common ancestor. At every point on the tree, organisms were successful enough to leave offspring. The tree of life is therefore a tree of unbroken success.

A more provocative way of thinking about adaptation is to say that living things are well fitted to their surroundings because they have evolutionarily formulated and tested an adequate hypothesis about their situation. Such an adaptive hypothesis consists of the totality of form, physiology, and behavior of an organism. It incorporates information about potential causes of failure and death—competition, predation, disease, and weather—as well as potential opportunities such as food or mates into a material body, a metabolizing "survival machine."

The concept of an adaptation as being equivalent to a scientific hypothesis, first proposed by the Austrian marine biologist Wolfgang Sterrer, introduces three core ideas of this book. First, a living thing is an adapted whole, all of whose characteristics affect its fate in the struggle for life. Natural selection due to enemies and allies operates on wholes, not on genes or on isolated traits encoded by those genes. Selection may change the frequency of gene variants controlling such traits as eye color, blood type, or susceptibility to diseases like hemophilia, but it is the organism's phenotype—its form, physiology, and behavior—that is exposed to selection and that constitutes the adaptive hypothesis. Our fellow meat-eaters on the Pliocene plains of East Africa competed with whole people, not with our adapted parts or our genes. Their fates and ours hinged on how well the bodies of competing individuals worked.

This idea is most obvious when applied to individual organisms, but it applies also to cells within the living body. In the nervous system, for example, axons—long extensions of nerve cells—grow more or less at random, but they persist only when they reach their target as indicated by an electrical signal, that is, when a connection is established. Other axons wither away, leaving an orderly, flexible, and well-functioning network of nerve connections that allows an animal to sense and respond to its environment. This trial-by-success pattern of axon growth and death yields an adapted whole in the same way that natural selection does at the higher level of biological organization: that of individual organisms.

The notion of an organism as hypothesis also points to the role that living things play in modifying their environment. Organisms do not simply respond physiologically or behaviorally to their surroundings, or evolve through natural selection in response to a new situation; they also help create their environment. The good fit between organism and environment is a combination of acting and being acted upon, a feedback rather than a one-way relationship. Organisms are active participants in their adaptation, not passive players unable to influence their own fates.

Third, the trial-by-success procedure that lies at the heart of adaptation in nature is essentially identical to the human scientific way of knowing. In nature, feedbacks between body and environment impose selection and yield adaptive wholes by distinguishing between those that survive and those that do not. In the scientific search for objective reality, a scientist consciously uses observations and prior evidence to formulate a hypothesis, which is then tested and modified with additional evidence gathered through further observation and experiment. Evolutionary adaptation is achieved without a guiding sentient mind, whereas science for the most part proceeds deliberately and purposefully; but the two adaptation-producing methods have in common a feedback between the environment and a meaningful, predictive representation of that environment.

To see this parallel between adaptation and science in more concrete terms, let's return to the plants and animals on the coast of Maine. The organisms Edith and I met on Squirrel Island all belong to species that are native to the North Atlantic. It therefore seemed reasonable to conclude that their distinguishing features, including their adaptations, evolved in that ocean, and to assume that the dangers and opportunities important to individual members of species on these shores are representative of conditions faced by each species as a whole both now and in the past. These ideas were my preliminary hypotheses about the origins of the adaptive characteristics that these common species displayed so clearly.

But when I began to examine these intuitions, I quickly discovered that my hypotheses would require substantial revision. For example, it became clear that the environments in which we observed our seashore species were far different from those that existed four hundred years ago and earlier, when many of these same species were flourishing. In those few centuries, humans had greatly altered the complement of predators, competitors, and food sources in the waters of New England and southeastern Canada. Gone were the vast numbers of lobsters and cod—major predators of clams and sea urchins—that French fishermen and early English settlers encountered in the seventeenth century. Three species—the great auk, Labrador duck, and sea mink—that had once been significant consumers of shore animals and fish had disappeared before 1920. The Atlantic gray whale, a voracious bottom-feeding species, had been hunted to extinction by 1675. Meanwhile, species from elsewhere arrived on these shores, including a rockweed, the common periwinkle, and the green crab, all accidentally introduced with ships' ballast rocks from Europe before 1840. The adaptations displayed by the native shore species in Maine clearly had been honed under conditions quite different from those that prevail in the present day. More refined hypotheses about the nature and origins of these adaptations were clearly needed.

To pinpoint the origins of New England's marine life and its adaptations, I needed to trace lineages back through millions of years of evolutionary time. The categories—families and genera—in which our New England species were classified are found in other geographic areas as well, which means that an adequate hypothesis of evolutionary origin and change would have to consider many species besides those found in Maine. Our species might have arisen in places far away and in environments that no longer existed. Their adaptive hypotheses might work in the human-altered conditions prevailing today, but they were evolutionarily formulated and honed elsewhere.

These possibilities began to engage my attention fifteen years later on the other side of the continent. Edith (by then my wife of fourteen years), our four-year-old daughter Hermine, and I had settled in for a spring sabbatical of research and teaching at the Friday Harbor Laboratories on San Juan Island in the state of Washington. Friday Harbor is an idyllic place, combining a rich and easily accessible marine fauna, a forest preserve perfect for long morning walks, and a stimulating intellectual atmosphere. Contemplating the specimens I collected during our stay, I realized that many of Friday Harbor's seaweeds and marine animals have close counterparts in the biologically much more impoverished floras and faunas of the North Atlantic. Even though the Pacific and Atlantic species were now separated by the impenetrable ice barrier in the Arctic Ocean, they must have had a common origin. But where was this place of origin, and what were conditions like when the adaptations of these species were evolving?

I settled on two hypotheses. Either the common ancestors of Atlantic-Pacific species pairs originated in the North Pacific and spread northward through the Arctic Ocean to the Atlantic, or they originated somewhere in the North Atlantic and moved through the Arctic into the Pacific. Either way, the spread of a temperate-zone ancestor from one northern ocean to another required a much warmer Arctic Ocean than the ice-covered ocean of the last eight hundred thousand years. Furthermore, the land bridge that linked Alaska and Siberia for tens of millions of years would have to have been breached by the sea to form the Bering Strait, which connects the North Pacific and Arctic Oceans. Testing and refining these hypotheses would involve gathering evidence from many sources. And so began a research project that has engaged me off and on for more than twenty years.

One obvious source of evidence, especially to a paleontologist like me, is the fossil record. Happily for me, seashells are abundant as fossils, and most species of living clams and snails have a fossil record that can be traced back for millions of years. If the lineage leading up to a living species extends further back in time in the Pacific than in the Arctic or Atlantic Oceans, it likely originated in the Pacific. Using genetic characteristics that can be observed in fossils, such as details of shell form, paleontologists can link ancestors with descendants. With the advent of molecular techniques, in which DNA sequences of related species are compared, it has become much easier to identify species lineages and to determine the geographic locations and times of lineage splitting. Geological evidence can help to specify when the Bering Strait was open, and when the Arctic Ocean was warm or when it was covered by sea ice.

Some of the evidence was already available through the work of other scientists. On my regular visits to the Smithsonian Institution's National Museum of Natural History in Washington, I combed through the literature on the living and fossil life of the northern oceans. There were hundreds of papers and monographs on which I took extensive notes. I could easily cope with the English, French, and German publications, but it wasn't until Janice Cooper came to work with me in Davis in 1989 that I came to appreciate the richness of the extensive Russian work, much of which Janice was able to translate into English.

The project called for a great deal of original work as well. I examined and compared thousands of specimens at the Smithsonian, home of the largest collection of shells in the world. Row upon row of cabinets, their drawers stuffed with trays of carefully labeled and documented specimens, fill a large part of the second floor of the east wing of the museum. Every day I spent in that collection and in its accompanying library yielded surprises—Arctic whelks with thick, ornate shells; delicate deep-water clams from the Bering Sea; and perfectly straight razor clams more than eight inches long from Scotland—and a deepening appreciation of museums as sanctuaries for the preservation and study of the world's biological treasures.

For molecular work, I turned to Tim Collins, then a postdoctoral student at the University of Michigan. We decided to concentrate on dog whelks, common predatory snails found throughout the northern hemisphere. Eight or nine species of dog whelk, belonging to the genus *Nucella,* live in the North Pacific, and only a single species—the one Edith and I observed at Squirrel Island—is known from the Atlantic. Tim already had specimens of the Atlantic species and of those from the northeastern Pacific, and we supplied specimens from the Aleutian Islands and northern Japan. The long fossil record of *Nucella* in the Pacific, extending back to at least twenty-five million years ago in California, already suggested that dog whelks originated on the American side of the Pacific, moved westward to Japan about sixteen to seventeen million years ago, and spread to the North Atlantic via the Arctic beginning about three million years ago. The evolutionary relationships of the living species inferred from Tim's molecular work confirmed this historical hypothesis.

To gain insights into the North Pacific environments from which the ancestors of Atlantic species came, I wanted to visit places besides the San Juan Islands. The frigid shores of the North Pacific were not an especially inviting destination for someone like me, who is most at home in warm tropical seas, but curiosity overcame any reservations I might have had about going. A once-in-a-lifetime opportunity presented itself in June 1987. Along with my good friend A. Richard Palmer, a brilliant experimental biologist at the University of Alberta, I was invited to take part in a research cruise aboard the *Alpha Helix,* a notoriously unstable but otherwise comfortable ship operated for the National Science Foundation by the University of Alaska. A team of marine biologists led by Jim Estes of the National Marine Fisheries Service in Santa Cruz was conducting experiments and collecting data on sea otters and their food along the entire one-thousand-mile chain of the Aleutian Islands. While Jim and his fellow divers worked offshore, Rich and I were motored to shore in a small boat. Clad in bright orange survival suits in case we should fall into the ice-cold water, we made our way gingerly over the islands' shores, careful not to slip on the thick carpets of seaweed or to cut ourselves on large sharp barnacles. Hands and fingers numb quickly in waters as cold as 5 degrees Celsius, so I donned fingerless mittens in order to make observations and collect specimens for later study. Except for the fragrant flowers of sea peas, the land vegetation—a springy cushion of grass and cow parsnip, but no trees—gave off little of the sweet smell I had come to associate with northern meadows. With bald eagles squealing overhead and Steller's sea lions barking from rocks nearby, there was no doubt that we were in a wonderfully remote outpost, visited by fishermen but no longer permanently occupied by civilians.

There is no substitute for experiencing an environment firsthand. Before our expedition, I had read extensively about the Aleutians, but nothing prepared me for the extraordinary luxuriance of the seaweed meadows clothing the rocks along the shore or for the scarcity of crabs and sea stars, predators that voraciously consume mussels, snails, and barnacles farther south both on the North American and Asian sides of the North Pacific. By the standards of these more southern sites in Washington, British Columbia, and the northern Japanese island of Hokkaido, which Edith and I visited one year later, the shells of Aleutian shore snails were conspicuously thin and delicate, apparently reflecting the minor role that predators played in the lives and evolution of these animals. The specimens that Jim Estes and his fellow divers brought up, by contrast, were often thick and robust, consistent with the presence and effects of predatory sea otters. If the Pliocene environments near Bering Strait, located thirteen to fourteen degrees of latitude north of the Aleutians, were anything like the shores we visited along that island chain, the ancestors of the tidal species that spread into the warm Arctic and Atlantic beginning some three-and-a-half million years ago would have been adapted to a regime of abundant food and little interference from predators, whereas species that lived below the tidal zone would have included intense predation as part of their adaptive hypothesis.

The historical hypothesis that has emerged is an explanation involving migration, climate change, and adaptation. For a period of three or four million years following the initial opening of the Bering Strait about 5.3 to 5.4 million years ago, the coastal Arctic Ocean seems to have been ice-free in summer or perhaps even year-round, allowing hundreds of species—fish, seals, crabs, sea stars, sea urchins, seaweeds, clams, and snails—to spread from one temperate ocean to the other. The number of lineages extending their geographic ranges from the Pacific to the Atlantic outnumbered those expanding in the opposite direction ten to one. The few Atlantic species that colonized the Pacific, including a handful of small clams as well as members of the herring and cod families, did so soon after a marine connection between the Arctic and Pacific Oceans was established, perhaps because the predominant flow of ocean currents through the newly formed Bering Strait was from north to south rather than from south to north as it is today. The great invasion from the Pacific to the North Atlantic began during the warm mid-Pliocene, about 3.5 million years ago, perhaps slightly after the first wave of climate-related extinction of native North Atlantic species. The adaptive hypotheses formulated by organisms about Pacific environments were successfully transferred to the Atlantic, where they were subtly modified according to the novel conditions encountered there by the newcomers. Further adaptive tweaking must have taken place as glaciers advanced and climates cooled beginning about 2.7 million years ago and as genetic contacts between Pacific and Atlantic populations were severed in the increasingly ice-covered and unproductive Arctic Ocean.

Like the adaptations of the species I was studying, my initial hypothesis about the origins and environments of the ancestors of Maine's seashore species had evolved. The naive notion that present-day causes of natural selection suffice for explaining adaptation was replaced by hypotheses that took into account the facts of geological and evolutionary history; and even these hypotheses have undergone change. When I began publishing papers on the exchange of species between the Atlantic and Pacific, the available evidence indicated that the Bering Strait opened about 3.5 million years ago, at the same time that the great influx of Pacific species in the Atlantic got underway. Detailed geological work in Alaska and on the Kamchatka Peninsula of Siberia has since shown that the Strait opened earlier, 5.3 to 5.4 million years ago, as shown by the arrival of a few Atlantic clams in the North Pacific at that time. Scientific hypotheses, like adaptations, are provisional, always subject to refinement or modification as circumstances warrant.

The similarity between evolutionary adaptation and conceptual evolution, as represented by the scientific way of knowing, is not merely a matter of analogy or metaphor. It represents deep homology, a common causal chain of feedbacks between living things and their surroundings. For organisms, the process entails detecting, acting on, and being affected by the environment through a combination of metabolic activity and natural selection. For scientists, it begins with careful observation and description, often expressed in mathematical terms and summarized with a model. But description is only the first step toward explanation, which is a hypothesis about how something works and how it came to be. This second step requires an understanding of causal agencies, of mechanisms and feedbacks between causes and consequences. To arrive at a scientific explanation, scientists must conduct experiments and make new observations and connections. They must manipulate old information and acquire new data. The final step is prediction. The adaptations of organisms are the material expression of an evolved response to the predictable elements of their world. For the scientist, as for the evolving organism, prediction involves extrapolation—applying an explanation that works for known circumstances to similar future conditions.

Nature, it can be said, invented the scientific method long before Aristotle first observed the nonhuman world systematically and long before European scholars in the twelfth century began to codify scientific inquiry. The sequence describing the evolutionary, or adaptive, way of knowing came into being three-and-a-half billion years ago when life first emerged on our planet. Life is set apart from nonlife by the establishment of causal links among structure (form, physiology, and activity), function, and fate (survival and propagation, or restriction or death). Science mirrors living nature by applying the same trial-by-success methods to uncover nature's secrets that organisms have employed through adaptation to build a robust biosphere of design and creativity.

It is hard to overestimate the power and reach of the evolutionary and scientific way of knowing. Thanks to adaptation, life has thrived and blossomed in a vast profusion of diversity for billions of years under widely varying conditions, including catastrophic events that brought about mass extinction but not the total elimination of life. Through adaptation, living things have adequately predicted and often improved upon their environments; they and the evolutionary lineages to which they belong are successful because their performance is under constant scrutiny by the agents of natural selection. Likewise, science is the most successful method for gaining human knowledge because its concepts are being constantly tested and improved through the relentless search for verifiable evidence. Theories, such as the theory of evolution, represent our best efforts to organize, explain, and predict what we know and what we don't yet know. As coherent bodies of knowledge linking disparate observations and hypotheses, theories are not just hunches or speculations that evaporate like mist in the morning sun; they constitute a durable framework, solid enough in their construction to withstand remodeling without falling to pieces yet flexible enough to accommodate additions and renovations. At the age of more than 150 years, the theory of evolution has changed dramatically since Darwin and Wallace first sketched it out in 1858, but its foundations are firm and its explanatory power is second to none.

As powerful as biological adaptation and the scientific way of knowing are, neither is infallible. An adapted organism's hypothesis fails when its environment is affected by an unpredictable catastrophe such as the collision of Earth with a comet or asteroid or the arrival of a new potent predator on an island like Hawaii where no such predator had existed before. Sometimes a new, better hypothesis comes along to replace the old even if other conditions remain unchanged. The predatory marsupials (pouched mammals) that were the top consumers in South America for millions of years could not match the more powerful dogs, cats, bears, and weasels that entered South America from North America when a land connection in Central America was established about three million years ago. Adaptation is as good as it has to be under the conditions at hand, but it cannot anticipate all possible future states.

Science and scientists, too, can fall short. They are vulnerable to the political and social currents of the day, and many good ideas are ignored either because their creators are seen as outsiders or because the ideas challenge assumptions, techniques, and theories on which the careers of prominent scientists and policy-makers are built. When Harvard biologist Edward O. Wilson published his book *Sociobiology* in 1975, in which he expansively laid out his thesis that much of behavior, including human social behavior, is underlain by genes, three leading evolutionary biologists—Harvard's Stephen Jay Gould and Richard Lewontin and the University of Chicago's Lee Van Valen—savagely attacked him in the journal *Nature* and elsewhere for fomenting ideas that seemed to support discriminatory policies on the basis of race and that were at odds with their own Marxist leanings. The resulting furor made the new field of sociobiology so unpopular and its practitioners so reviled that the field was renamed "evolutionary psychology" in order to distance itself from Wilson. It is likely that Wilson overstated his case, leaving little room for direct environmental effects on the development of the brain and the rest of the nervous system; but abundant evidence linking behavior with genetic markers that run in families has not quelled the deep suspicion many scientists still hold against this important discipline.

Like other people, scientists make mistakes. They confuse description with explanation, draw incorrect or unsubstantiated conclusions from the evidence, make unwarranted assumptions, misapply or misinterpret statistics, and in rare cases lie, cheat, and steal. Fortunately, as the University of Wisconsin philosopher David Hull has convincingly argued in his investigations of how science and scientists work, the interests of individual researchers and those of society at large usually coincide, so that the ethic of accuracy and honesty generally prevails. Moreover, most scientific work is carefully scrutinized by other scientists, so that flaws are often quickly spotted and eventually corrected. As an inherently social activity, science is thus self-correcting thanks to the link between the rewards to individuals—reputation, status, and money—and the benefits to society, which include better understanding, technical innovation, and the triumph of free inquiry.

Those who are skeptical of the parallel between the methods of science and the process of evolutionary adaptation might argue that, whereas science and its public acceptance are social phenomena, adaptation in the nonhuman domain of nature is more firmly situated at the level of individual organisms. I believe this supposed contrast is greatly exaggerated. Although science is often a collaborative enterprise, individual researchers play decisive roles at every stage in the formation of scientific concepts, from making observations to the proposal of grand theories. Darwin and Wallace were instrumental in laying the foundations for the theory of evolution, and although the theory would undoubtedly have triumphed without them, other individual architects would have had to take their place. Theories are spawned by creative individuals and honed by thousands of other scientists working alone or in collaborative groups. Science proceeds in a social milieu, but as long as individuals are rewarded for doing good science, they will continue to exercise a great deal of control over what science accomplishes and how it is conducted.

Well-adapted organisms are often thought of as independent, self-sustaining individuals, but their features are fashioned in an ecosystem, a community of species that provides food and supports enemies. Other individuals determine which attributes work and which do not. Hypotheses, whether they be scientific or evolutionary, are embedded in a network of individuals who are competing and cooperating with one another in an economic world of costs, benefits, opportunities, risks, and challenges.

Another alleged difference between the scientific way of knowing and the process of evolutionary adaptation is that science appears to be progressive, whereas evolution seems not to be. It can hardly be doubted that science has followed a general upward trajectory toward a more accurate, more comprehensive approximation to empirical truth, and that it has resulted in an overall, though not universal, rise in the human standard of living. The goals of science are consistently pursued, and scientific knowledge is cumulative. By contrast, evolutionary adaptations have been portrayed by late Harvard biologist

Stephen Jay Gould as local and ephemeral. According to this view, there is no "goal" toward which adapted organisms collectively "strive." This interpretation has led Gould to criticize views held by the British evolutionary biologist and philosopher Julian Huxley and many others that evolution is progressive because it results in such humanlike attributes as complexity and the ability to do science. Words like *primitive* and *advanced* were common currency in writings about evolution, and Gould rightly rejected such language. The Western notion of progress—a trend toward greater wealth, justice, friendship, and freedom—does indeed seem to be a poor metaphor to use in the history of life on Earth, or arguably about much of human history. As Gould and many others have pointed out, there are far more examples of evolutionary lineages tending toward a simplification of structure than there are lineages that have become more complex. I therefore have no quarrel with Gould when he writes that evolution is "not the conventional tale of steadily increasing excellence, excellence, complexity, and diversity."[1]

But there is another way of looking at evolutionary history. Adaptation is the way in which living things make sense of the world—it enables unusual phenomena related to resources, weather, and enemies to be incorporated into a predictive hypothesis, which is retained, transmitted, and refined in subsequent generations. Adaptive "knowledge" is in this sense cumulative, especially because many adaptations work well not just in one particular situation, but in a wide range of conditions. Moreover, the most influential adaptations—those that enable their bearers to increase their own opportunities and to control the lives of other species—repeatedly and predictably arise. In the earliest phases of life's history, the resources on which living things depended were controlled by inanimate processes operating on the Earth and around our young sun. The cumulative adaptations of organisms, however, established feedbacks between living things and their resources, and unintentionally but inexorably created a system of regulation where the essential nutrients and the ingredients of living bodies came under the control of life itself. The competition for locally scarce resources that enables winners to acquire more resources and to capitalize on more opportunities yields life-forms with greater power, and these agents—top consumers and major producers in their ecosystem—create additional opportunities and positive feedbacks with resources. Although these relationships are interrupted from time to time by environmental disasters beyond life's immediate control, there is a cumulative reinforcement between living things and the resources needed to sustain them. Like humans employing the scientific method to push back ignorance, life through its adaptations is pushing back the realm of unpredictability by fashioning its own predictable environment and by expanding the range of adaptive options. The evolutionary interplay between life and environment is the subject of much of the rest of this book.

Two other differences deserve brief mention here but will be dealt with more fully in chapter 6. In nature, adaptive characteristics are usually transmitted, or inherited, by genes, whereas scientific advances are made and transmitted culturally. This difference is less stark than one might think. Not only are many vertically inherited traits—those passing from one generation to the next—under substantial nongenetic control, indicating a direct role for physical forces and chemical conditions in determining an organism's phenotype, but many genes can move horizontally from one organism to another in the same generation. Biological adaptation therefore has much in common with the way scientific ideas and other cultural features form and spread. Second, as already noted, science proceeds deliberately, whereas adaptation takes place without conscious intentionality. The main consequence of this difference is that the adaptive process takes place much faster in the minds of individual humans than it does across generations of organisms in nature. We can alter our ideas in an evolutionary instant, whereas gene-based adaptive change may take years to decades. Humans have therefore made a huge advancement in the speed of adaptation, and have made the whole method far more flexible; but the underlying process is not materially different from adaptation in the realm of nonhuman creatures.

In short, science and the evolutionary way of knowing are based on the same trial-by-success method. They may differ in speed, in intentionality, and in the means of transmitting accumulated "knowledge," but they have both proven to be robust, successful, and flexible pathways toward solutions that work in a changing and challenging world.

If the scientific endeavor is so successful and so generally advantageous to society, why is it that worldviews founded on doctrines that must be accepted without question or evidence have also been so successful throughout human history? In the language of adaptation, how can such doctrines, which often clash irreconcilably with the facts and theories as revealed by scientific inquiry, be construed as adaptations, or as hypotheses of the human environment? How, in other words, do irrationality and knowledge by decree, which contradict the adaptive process central to both evolution and science, fit into an explanatory framework based on the principles of adaptation in a finite world?

The resolution of this paradox is to be found in our social nature and in the way that humans achieve group identity. Throughout the animal world, groups often hold substantial advantages over individuals. They compete more effectively for resources, offer better defense against predators, and are able to perform many essential tasks at the same time. Groups work well only if they are held together by a common bond that transcends individual self-interest. In insect societies such as those of ants and termites, all individuals are closely genetically related to one another, so there is little conflict between individual interests and the common good. Individuals in human groups, by contrast, are genetically distinct. Group cohesion in our species must therefore be based on a common set of beliefs, rules, and customs. The cultural foundation of human groups is expressed in doctrines of allegiance, manifested as religion, nationalism, team spirit, and the like. These doctrines, which often exact a significant cost to individuals, are highly effective human social adaptations, especially if they are enforced vigorously by leaders or their surrogates. We need authorities to keep renegade individuals in line and to ensure that the doctrines holding groups together do not dissolve. Shared mythologies, whether they are based on deities, charismatic leaders, romanticized or glorified stories of the past, or a stable core of secular laws, are effective hypotheses of the human social environment because they work to unite individuals of different backgrounds and interests, even if these cultural traditions contradict scientifically acquired knowledge. The criterion of their success is not how well they approach objective empirical truth, but how well they enable stable groups to function in the context of competition with other groups.

The paradox of coexisting rational and irrational styles of thought is resolved when we consider the criterion of success of adaptive hypotheses. David Sloan Wilson, an evolutionary biologist at the State University of New York at Binghamton, puts it this way: "Rationality is not the gold standard against which all other forms of thought are to be judged. Adaptation is the gold standard against which rationality must be judged, along with all other forms of thought."[2] Whether we like it or not, it is what works that matters, not always what is verifiably true. The hypotheses that organisms and science reveal are approximations to evidence-based reality. They work because such reality is crucial to survival and propagation. Social adaptations such as doctrines of allegiance may bear no resemblance to verifiable truth, but they work because they profoundly influence human lives. Cultural bonds like religion and nationalism, which are accepted on authority rather than on scientific scrutiny, will be with us as long as there are human groups that compete for a place in the sun.

These same social forces are at work when inadequate or incorrect empirical theories persist, despite evidence that other explanations better account for the known facts. The idea that the continents could move over the course of geological time across Earth's surface was resisted for decades by American geologists despite strong circumstantial evidence from multiple sources that Africa and South America had separated to form the South Atlantic Ocean beginning some 135 million years ago. As a social process in which ideas spread not only because of their explanatory power but also because influential people champion them, science cannot escape the social forces that sometimes—perhaps often—have a greater influence on the fate of ideas than does the empirical validity of those idea. Success is bound up not with truth alone, but with social acceptance, with prevailing cultural norms, and with the status of people for and against a doctrine, scientific or otherwise.

For those like me who reject doctrines based on unknowable, supernatural beings or on unquestioned beliefs imposed by totalitarians, the question is not whether we can rid ourselves of such fictions. The historical and anthropological evidence overwhelmingly indicates that such beliefs, which appeal to people's emotions and which serve to differentiate groups, appear in all known human societies. Instead, the question is whether it is possible to construct a socially cohesive system of thought that not only enhances the common good, but is founded on and compatible with reality. With the scientific and evolutionary way of knowing taking an increasingly dominant role in human society, we need a better adaptive hypothesis of our social environment, a hypothesis that eliminates contradictions with evidence-based findings and that at the same time incorporates a powerful moral contract among group members. Like all adaptations, such a contract must be seen to benefit individuals as well as the groups to which they belong. They must preserve flexibility and the capacity to change as social circumstances warrant. I shall return to this social dimension and its implications for the emergence of meaning and purpose in chapter 8.

Adaptations do not come ready-made. They arise as imperfect solutions to problems and opportunities and are then refined by selection. The entire process inextricably links organism and environment, which interact and influence each other. How, then, does it all begin? Back in the tidepools and kelp thickets on a Maine shore, how do the organisms that live there discover what is dangerous? How do they incorporate environmental information to their own ends? How, in other words, do they acquire the evidence necessary to formulate an adequate evolutionary hypothesis of their situation? And how do we humans—scientists included—begin to make sense of the world? It is to these questions that I turn in the next chapter.

CHAPTER TWO

Deciphering Nature's Codebook

What would it be like to be thrust into a completely alien world? How would one adapt to such a situation, in which the surroundings reveal no discernible information and where, as a result, no basis for an appropriate response exists? Such a possibility may seem preposterous, relegated to the realm of science fiction; but this is precisely the predicament that early life-forms confronted and in which human babies find themselves. Newborn babies have no language, no self-awareness, and only an all-purpose cry for communicating their needs. A chaotic jumble of meaningless signals must somehow be transformed, organized, and recorded so that meaning—something that makes sense and about which a hypothesis can be formulated—emerges. Some form of perception, of environmental information transmitted to and interpreted by an organism, seems to be essential for the process of adaptation, the formation and testing of hypotheses about the world. This is the aspect of adaptation, together with the implications for the education of children, that I explore in this chapter.

Immigrants have some inkling of the challenges of a wholly unfamiliar world. As one of those immigrants, I well remember my bewilderment when, as a boy of nine, I moved with my family from the Netherlands to an isolated, rundown farm in the hills of northwestern New Jersey. With just a few weeks of rudimentary introduction to English at school in the Netherlands, I barely understood the language and certainly could not speak it. Foods—breakfast cereals, concord grapes, persimmons, and pumpkin pie—at first seemed strange, and bread was peculiarly sweet. The forests in the area were exciting, wild, forbidding places where dangers I had not known back home—snakes, bears, poison ivy, and venomous spiders—could strike at anytime. The Dutch pine woods and polders—lush grasslands lying below sea level—seemed safe by comparison. I fancied myself as one of those European explorers of the late fifteenth and sixteenth centuries as they ventured into unknown lands.

I rapidly learned to decode this new world and to make my way in it. With the help of dedicated and patient teachers, I learned English, mastered contracted English Braille, and more or less successfully integrated into the chaotic milieu of the third-grade class in Newton Elementary School. But I never forgot that all this adaptation to American life hinged upon acquiring and extracting meaning from information that came at me in a steady stream of signals.

Whether one is a baby, an immigrant, a scientist, or a primordial organism, the challenges of extracting information from the environment and formulating a hypothesis about the predictable components of that environment are always the same. Signals from the surroundings must be sensed by a receptor, a molecule or larger unit that is consistently affected by a given energy source—light, heat, a magnetic field, an electric charge, a mechanical force, or a change in the concentration of a substance—and must be recorded and transmitted to sites where they are culled and organized to form an emergent, integrated representation or pattern consisting of measurements and comparisons. The organism must then have the capacity to respond, either passively or actively, through a locally or centrally situated control device. Adaptation—the formulation of a hypothesis about the world—is therefore an emergent process and outcome, founded on sensation (or observation), selective culling (or evaluation), and coordination and collaboration among receptors and responders, all orchestrated by and for the emergent goals of survival and propagation, and all made possible by tamed energy emanating from outside and inside a living, metabolizing body.

It is that first step—detection, observation, and measurement—on which adaptation and the scientific way of knowing depend. Without signals that we can sense, we know nothing. There is no basis of prediction, no grounds for a hypothesis, no capacity for adaptation. Danger is indistinguishable from opportunity, and an organism's structure can neither track nor reflect conditions outside or inside its body. A signal by itself is meaningless. In order to respond to it, we must know what it represents; we must distinguish among signals, and there must be some way of ascertaining a signal's intensity and direction. If the information gathered is not recorded or retained in some form, it cannot accumulate or be applied. A memory, whether encoded in genetic material or stored in books, on computers, or in more ephemeral molecular form in the brain, is therefore an essential ingredient in sense-based adaptation.

The world is replete with potentially informative signals, so full in fact that we—and all other life-forms—would be completely overwhelmed if we sensed them all. It is the structure of pigments, proteins, and cell membranes that determines which signals can be detected and how information that is meaningful is conveyed. The lipid molecules that make up most of the cell membrane are highly sensitive to electrical signals, which form the basis of much cell to cell communication and of the nervous system. Ion channels—gateways through the cell membrane that enable calcium, potassium, and sodium ions to pass into and out of the cell—allow the cell to carry out internal functions and to interact economically with the outside.

Sodium transport, which permits the reception and transmission of nerve impulses, functions in bacteria in respiration and in the rotational motion of flagella, structures that enable many bacteria to move from place to place. The ability to transport sodium evolved from an already existing capacity to regulate calcium, which has innumerable functions including the control of genes, cell division, and the movements of components within the cell. Acetylcholine, which transmits an electrical potential from a nerve cell to a muscle cell across a narrow gap (synapse), occurs widely in bacteria, plants, and other organisms without a nervous system, where it is essential in communication between cells. In land plants, proteins known as phytochromes are sensitive to long wave-length far-red and green light. It is at these frequencies of light that plants detect, and respond to, shade cast by neighbors. Selective molecular detectors and transmitters thus occur widely in living things, and form the basis of all sensation and adaptation.

The basic building blocks of detection and communication set the stage for dramatic elaboration in many animals. They work because sources of energy change the shapes, positions, and often the charge of the molecules involved. Living things capitalize on this universal phenomenon of energy being captured and stored by molecules. Elaboration of these simple molecular sensors and memory devices is strongly favored because it improves the ability of organisms to identify food, danger, and other aspects of the environment that affect survival.

Consider, for example, the evolution of animal eyes. The simplest eye consists of a single light-receptor cell and a shading pigment cell. Such eyespots cannot form images as more sophisticated eyes do, but they can detect the direction of light. In the plankton-dwelling larva of a worm studied by the Hungarian cell biologist Gáspár Jékely and his colleagues, the arrival of light at the eyespot stimulates a nerve, which changes the beating of nearby cilia, redirecting the currents created by these cilia in such a way that the larva swims up toward the light. Even in this simplest of systems, therefore, there is coordination between a sensor and a motor response.

Selection imposed by competitors has caused living things to build on the innate receptiveness of proteins and pigments by elaborating sense organs, nervous systems, and memory devices. Light-sensitive proteins known as rhodopsins are found in bacteria, single-celled "protozoans," and animals; but sophisticated eyes based on rhodopsins have developed in some forty to sixty separate groups of animals, which use them for detecting danger, food, and mates. The detection of electricity, which appears to be intrinsic to nearly all cells, has been carried to great heights in many fishes, including sharks, rays, catfish, and electric eels. In conjunction with underwater hearing, these fish use electrical signals at night or in cloudy water to find and attract mates, locate food, and become aware of predators whose movements produce detectable electrical impulses. Army ants use pheromones to communicate with one another and to coordinate the activities of individual workers so that the colony effectively acts as a single, fearsome predatory machine. These examples show that the simple detectors that might suffice in the small-scale world of tiny, slow-moving life-forms no longer make the grade when larger, high-speed competitors capable of sensing and influencing conditions many body lengths away from them enter the picture. Predators that eat widely scattered prey, and must therefore travel large distances, need to have particularly keen senses.

There is remarkably little comparative information on the sensory performance of species, but my observations and a close reading of the technical literature indicate to me that the most highly developed senses of hearing, echolocation, smell, vision, and electric perception—all connected with detecting objects at a distance—are found in animals that hunt highly mobile prey. The most sophisticated eyes occur in predatory squid, raptorial birds like hawks and eagles, marine mantis shrimps, jumping spiders, predaceous insects including roving beetles and praying mantises, fossil marine reptiles known as ichthyosaurs, and—surprisingly, given the absence of a central nervous system—lethally venomous cube jellyfish. Predators hunting at night or in turbid waters are famous for a keen sense of smell (or detection of chemical cues). Rattlesnakes in search of small mammals detect their victims by heat. Night-hunting bats and some birds, as well as whales swimming in the ocean depths, are like me in having developed a highly effective means of locating objects by transmitting bursts of sound, which then echo from these objects. All of the several fish groups with a well-developed electrical sense are predators, either in the open sea or in turbid or cluttered environments in rivers and on inshore sandflats and mudflats.

This suggests to me that selection for detecting resources at a distance may be even more potent than selection to sense predators. Snails illustrate this point well. Many snails that feed on seaweeds or scrape microscopic organisms from rock surfaces only react to predatory snails and sea stars after there is direct contact between the two parties. Snails with a carnivorous diet, by contrast, can detect such predators at a distance by sensing the direction and concentration of an attacker's odor plume with well-developed olfactory organs located at the front end of the animal.

Why might this be so? Two reasons, one concerning the prey and the other concerning the predator, together explain the sophisticated detective abilities of active hunters. First, highly mobile items of food tend to be widely scattered and are hard to locate. Moreover, they often possess well-developed senses that alert the prey to the presence of enemies. Successful identification of prey at a distance is therefore essential for a predator specializing in a diet of scarce, elusive animals. Second, there is intense competition among predators at every stage of the hunt, from recognition and capture to killing and eating the prey. A predator capable of discriminating suitable prey from unsuitable items at a distance will thus gain a good head start over its rivals.

Once long-distance detection has been acquired, the evolutionary door has been opened to other possibilities. For example, it allows for the evolution of means to detect and attract mates from afar. The imperative to find mates exists in all animals that practice internal fertilization as well as in those fish in which fertilization, though external (with union between egg and sperm taking place in the water outside the body of the female or male), occurs in a restricted territory. It even applies to flowering plants, in which the internal fertilization, or pollination, requires the help of mobile animals such as bees, butterflies, birds, and bats. It is intriguing that many groups of pollinators evolved from ancestors with predatory habits, raising the possibility that the ability to locate scarce prey at a distance predisposed some lineages to target widely scattered flowers as food sources and to enter into a pollination partnership with stationary plants that cannot fertilize one another on their own.

Many signals that are important in the lives of organisms come from other living things. Bees rely on ultraviolet marks on the flowers they pollinate. Many other pollinators home in on strong fragrances, or on colors in the visible part of the light spectrum. Pheromones, distinctive calls, visual displays, and other long-distance cues enable dispersed animals to find mates and to recognize members of their own species. Substances leaking into the surrounding water from an actively moving animal or a filtering clam guide predators to their prey. The mere act of swimming creates pressure changes—in other words, underwater noise—that swimming predators can detect. The immune system is able to identify and neutralize thousands of distinctive pathogens and toxins.

This quick survey of senses reveals a subtle but important evolutionary feedback between sensation and adaptation. The ability to sense the environment and to extract meaningful information from it is essential to the process of adaptation; but this ability itself becomes a target of adaptive improvement as competition for food and mates and the threat of predation increasingly influence the lives of individual organisms. The more we know, the more we need to know about our surroundings.

Long-distance communication among allies presents the obvious problem that attractive signals may also draw enemies. The sender must therefore be prepared to deal with those unwanted visitors, either by resisting attacks passively or by taking more active measures involving aggression or retaliation. If this problem is widespread, we should expect the most highly developed defenses to be found in species that use long-distance signals to attract mates, pollinators, and helpers. An effective alternative, employed by many noisy adult insects and by flowering plants, is to limit the conspicuous phase to a very short interval of time, just long enough to effect mating, fertilization, egg laying, or seed setting. In species that are not conspicuous to friends from afar, combat and passive resistance may be of lesser importance. Unfortunately, these predictions have not been thoroughly tested, but they illustrate how both the sender and the receiver of signals might adapt to the consequences of communication.

If organisms are important sources of signals to other living things, there is ample opportunity to transmit false or misleading information, or to render signals difficult to detect or to interpret. Lying and cheating obviously complicate the formulation of reliable hypotheses. Humans and many other animals form search images based on predictable shapes and behaviors of the objects for which they are looking. They therefore may fail to recognize desirable items that look, sound, smell, or feel different. Not surprisingly, therefore, inaccurate and dishonest signaling is widely, if not ubiquitously, employed by living things to deceive foes and to gain a competitive edge.

Palatable insects, for example, mimic dangerous or inedible ones.

The leaves of some passion-flower vines are spotted, looking to a female butterfly that is ready to lay eggs on the host plant as if the leaf were already covered with eggs. The host plant benefits by not having its leaves eaten by the caterpillars that would have hatched if the female butterfly had not been fooled. Viruses can deceive immune cells and colonize the body's cells without the usual immune defenses being mounted. Predatory spiders that chemically mimic ants can invade ant colonies without the knowledge of border guards because the ants accept them as kin. They are therefore at liberty to feed on the unsuspecting hosts. Advertisers, spy agencies, propagandists, criminals, and security officials will recognize these tactics as familiar and widely used means to influence people; but humans are hardly the first life-forms to exploit these methods or to benefit from deceitful communication.

If too many signals are dishonest or misleading, the link between sensation and adaptation would be severed, and the whole evolutionary project of adaptation would be compromised, stopped in its tracks. But the link still exists, and adaptation remains the universal property of living things that it has been ever since the first organisms populated the Earth three-and-a-half billion years ago. What, then, keeps deceit in check? How is it that the world of life is not overwhelmed with misinformation to the point where signals and patterns reflecting objective reality are unrecognizable?

Extensive research in human societies has shown that punishment discourages selfish cheating, and that rewards tend to encourage cooperative or truthful behavior. Modern societies attempt to prosecute false advertisers, and perjury—lying under oath—can land one in prison. These same incentives operate in nature. The nitrogen-fixing bacteria that live on the roots of peas and beans and that supply much-needed nitrogen to their hosts receive oxygen in return. The bacteria could become selfish by taking up oxygen and not delivering nitrogen, but the plant hosts police them by denying the bacteria oxygen if the supply of nitrogen is cut off. In ant and wasp societies, the queen is responsible for laying eggs and therefore exercises a reproductive monopoly. Workers, which are normally sterile females, sometimes lay eggs, but they are usually prevented from engaging in such selfish behavior by other workers.

More generally, however, dishonest signals work only if the receiver is momentarily misled, which is most likely when such signals are rare. If signals are usually unreliable, the receiver is apt to change its hypothesis of the meaning of the signals, with the result that the advantages accruing to the sender of falsehoods will disappear. Dishonest signaling must therefore always remain the prerogative of the minority, and honest signaling should remain the norm. Moreover, the existence of dishonest signals likely creates powerful selection favoring greater resolution, greater accuracy, and greater acuity in receptors and interpretative systems.

A signal achieves meaning when it is connected to a response. In the simplest case, a local signal is joined to a local response, as in the larval eyespot discussed earlier. At the next level of sophistication, signals are combined and compared, and noise is selectively removed to reveal patterns that can be mapped. A visual representation of the environment emerges when light signals are projected onto a richly innervated surface, such as the retina of vertebrates, cephalopods, and jumping spiders. The sense of touch likewise requires mapping of signals from nerve endings on an animal's body surfaces. Experiments by the Spanish paleobiologist Antonio Checa and his colleagues in Spain have shown, for example, that when a snail builds a shell, the shell-secreting mantle margin uses previously constructed parts as a guide, sensed through touch. The Princeton biologists James and Carol Gould, in their splendid book *Animal Architects,* describe many instances in which a nest-building bird or insect measures the size of the shelter being constructed by its own body. More detailed discrimination and location of objects in the environment, including those at some distance from the body, require more sophisticated mapping. In the case of sound and smell, intensity and direction of signals must be inferred from slight differences in the time of arrival at receiver sites. With these capacities, animals can navigate in the neighborhood for which they have created a multisensory representation. At the highest level of sensory sophistication, animals have an abstract concept of their environment. This enables them to modify and create structures or to execute other actions according to a preconceived goal or plan. According to the Goulds, this is the level achieved by beavers when these large rodents build, modify, and repair dams, which regulate the water level relative to the lodges in which beaver families live.

With each step in this gradation of sensory mapping comes an increased awareness and a corresponding ability to learn. There is an increasingly dynamic exchange between an animal and its environment, in which the environment in effect teaches an animal how to extract meaningful information from it. Very often, especially in social animals, learning involves a teacher, or at least a model, which is usually another animal.

Recent work on how language is learned underscores this point. A novice learns to associate initially meaningless sounds uttered by someone else with an object or an action when he or she observes the other person engaging with the object or in the activity while using consistent, different sounds. Language—the code of utterances that comes to be substituted for real things and events—is transmitted culturally. It evolves as initially naive speakers, having a necessarily incomplete grasp of the correspondences, predict regularities to fill in gaps. The resulting code therefore acquires not only meaning, but also a predictive structure of syntax and context, which can be further transmitted through imitative learning. Chance correspondences between sounds (or symbols) and concepts, things, and actions become stabilized and codified as language evolves. This concept of language evolution and learning, experimentally supported by the Scottish evolutionary linguist Simon Kirby and his colleagues, provides a nice model for how meaningless signals acquire meaning and, perhaps more provocatively still, for how the genetic code could have arisen and become essentially universal.

An incident in our daughter Hermine's English language illustrates these principles well. In the summer of 1983, when Hermine was one-and-a-half years old, we spent two months in New Haven, where I worked extensively in the libraries and museum collections at Yale in preparation for writing a book. In the large, overgrown garden of the house we rented, Hermine encountered a thistle and, being pricked by its spiny leaves, began to cry. "Thistle," I said, trying to explain what had hurt her. In Hermine's mind, *thistle*—which she pronounced "shishel"—became a noun meaning anything that caused pain, and also a verb meaning "to hurt." This initially broad association between pain and an object causing pain represented a reasonable hypothesis, which was, of course, subsequently modified to conform to the standard use of the word *thistle* as a noun to denote a particular type of spiny plant.

Another, more universal code that transmits and translates adaptive information in living things is the genetic code. This code, worked out by the English biologist Francis Crick and other molecular biologists in the 1950s, is based on a correspondence between the nucleotide components of RNA and DNA and the twenty common amino acids, the constituents of proteins. Each amino acid is associated with one or more sequences of three nucleotides. By themselves, such triplets have no meaning, but in the genetic code they unambiguously specify an amino acid, so that a string of triplets spells out the sequence of amino acids in a protein. Theories of the origin of the genetic code postulate an early phase in which related triplets corresponded to a group of chemically similar amino acids, with the result that there was a certain ambiguity in the specification of protein structure. A more precise correspondence evolved and stabilized when, by exchanging genetic information among cell lineages, only those lines that could "talk to" one another genetically could remain members of their communities. Ambiguous or deviant sequences were thus eliminated through intense selection. The common language of genetics has persisted with few minor variations through the realm of life.

For most living things, learning how to see, hear, smell, feel, taste, or detect electrical charge proceeds unconsciously and unintentionally through a kind of diffuse intelligence, in which many individual sensors act together in concert with other parts of the nervous system to produce a representation of the environment. Looked at in retrospect, the representation appears to be the product of directed intelligence, situated in a central authority like a brain or a deity. This perception, however, is an illusion, suggested to the human mind because conscious, intentional learning and exploration have come to predominate over the most diffuse methods seen in our progenitors. For humans, in other words, the ability to observe has become a skill, to be learned and honed through practice and intentional effort.

Despite its central importance as the starting point for the scientific—and evolutionary—way of knowing, observation as a learned skill has been neglected by the educational establishment. I experienced this neglect firsthand a few years ago when I was invited to take part in a summer program for blind children, eleven to fourteen years old, who showed some interest in science. Given my love of seashells, which I had been collecting and studying for decades, I decided to use these accessible and pleasing objects to introduce the children to the idea that scientific inquiry begins with observations and questions. Each child was handed a shell, which I then asked him or her to describe. After this first encounter, we could begin to probe the many oddities of form and ornamentation that these shells displayed so well. I hoped to duplicate for these children the wonder I literally felt when, at the age of ten, I first encountered shells from Florida and other exotic places. I would turn each shell over and over between my fingers, first perceiving its overall shape, then inspecting it more closely, running my fingertips over the surface again and again and using the ends of my fingernails to reveal the fine details. As my fingers moved over every accessible surface, I wondered why the ribbing on many of these shells was so regularly arranged, why the inner surfaces were so glossy-smooth, and why the range of shapes was so great compared to the limited variations among the North Sea shells I had known from the beach at Scheveningen.

I did not expect the children to notice the kinds of detail I look for, but I hoped they would take the time to inspect these objects carefully, given that few if any of the children would have handled anything like them previously. Each specimen expressed a wealth of tactile features, which cannot be grasped with a cursory inspection. Yet a cursory examination is all that these shells elicited. Adult observers in the room told me that it took about one second for a child to pick up the shell, examine it, and put it down without further touching. A bit of probing confirmed that these children had reached adolescence without having acquired the habit of automatic exploration through the sense of touch. Did this reflect a pervasive lack of curiosity? Had no teacher, parent, or sibling shown these children the pleasures and rewards of close tactile observation? Is there an unspoken assumption that gaining experience through touch comes naturally, especially to a person without sight? I came away from this experience convinced that observation, like reading and writing, is a skill that must be nurtured and honed before it becomes an unconscious habit of mind.

Neglect of curiosity and the ability to observe closely is by no means confined to the tactile sense of blind children. People with all their senses in working order often seem oblivious to their surroundings.

They do not hear the song sparrow or the house wren performing its coloratura from a nearby tree, or notice the fragrance of a citrus tree in bloom, or spot the ring of mushrooms that appeared overnight around a stump. Is it disinterest, or preoccupation with other more pressing concerns, that dulls our awareness of these nonhuman phenomena? Or have our senses and minds never been trained? Does the sensory environment of urban life and of human behavior simply overwhelm everything else? Why is the urban music that is piped directly into the ears of so many students preferable to the rest of the acoustic environment, even though it reduces awareness of potential danger? Is our alienation from nature and sometimes even from other people in part a consequence of not educating our senses?

During my formative years, my parents and brother acted as twenty-four-hour tour guides, never missing an opportunity to show me things within reach and describing things that were not. Many of my early teachers, especially in the Netherlands, likewise attended to the knowledge that can be gained by listening, touching, and even tasting the environment. Policy-makers and educators today seem far more interested in test scores and abstract measures of school performance. They are obsessed with perpetual evaluation of a very limited range of skills even as they ignore, or even discourage, curiosity and exploration.

But it need not be so, and indeed it must not remain so. If educators, businessmen, and politicians want children to learn about science, we must do more than teach mathematics or the facts and experimental methods of science. We must persuade educators—and ultimately ourselves—to place greater value on our senses and on their use in exploration with the thinking brain in gear. If training the senses is best done alone or in the company of one or two others, substantial blocks of time devoted to individualized attention will be needed to supplement and enrich the more usual, standards-based curriculum. Such a commitment will require more personnel and money, commodities that are already in chronically short supply in a profession to which society has rarely devoted enough resources. But hope springs eternal.

Editorializing aside, it is clear that exploration, enabled by multiple senses and multiple modalities, is a universal and necessary capacity of living things. It allows us all to read and decipher nature's codebook and to benefit from its contents. Misprints, confusion, misunderstandings, and deliberate deception abound, but all ancestors of life-forms now and in the past have read and comprehended nature's messages well enough to survive and reproduce. No living things—not even the most sophisticated human being with access to all the world's accumulated human knowledge—can read everything nature has to offer; and yet we mostly manage. Just how adaptation coexists with—indeed depends on—imperfection is the subject of the next chapter.

CHAPTER THREE

On Imperfection

When I first became attracted to seashells as objects of beauty and scientific study, I was like most collectors, valuing only perfect specimens. Something in me yearned for a kind of Platonic perfection, an ideal form in which all the characters of the shell were fully and faultlessly expressed. The mathematical regularity of the shell's spiral form, the flawlessly smooth interior, and the precise pattern of ribs on its outer surface spoke to me of a deep underlying harmony in nature's realm. Hopelessly naive and romantic, perhaps, but this imaginary world of unblemished beauty offered a welcome refuge from the often incomprehensible and bleak human world that lay beyond the safe confines of home and family.

But this was all to change as my intellectual horizons broadened far from home. The breakthrough came in the summer of 1970 during an early field excursion with my friend and colleague Lu Eldredge on one of Guam's windward reef flats. Splashing through hot pools left by the receding tide on a sun-drenched afternoon, we slowly made our way seaward in the direction of the reef crest, where the full force of the Pacific produced a menacing roar as the waves crashed into the complex topography of corals and rocklike coralline seaweeds at the reef edge. Stopping for a moment, Lu reached down to pick up a shell from one of the tide pools. "Look at this," he said, handing me a small broken shell. I had already been distressed to find so many damaged shells, so that I was at first disappointed to be presented with yet another. This was no ordinary ugly remnant of a once-intact shell. Lu had found a money cowrie whose shell was still so splendidly glossy-smooth that the snail inhabitant must have died just hours earlier. Marring the smooth contour of the shell was a huge jagged hole where the top of the shell should have been. The damage was so extensive that even a hermit crab desperate for a protective shell to house its soft abdomen would have rejected this object as uninhabitable. "Crabs in our aquarium do this," Lu continued. There were, Lu insisted, crabs with enormous claws that hid beneath boulders by day and hunted shell-encased prey by night. Curiosity kept me from flinging the specimen back into the pool. The cowrie joined several whole specimens in my collecting bag.

Damaged shells revealed truths that intact ones were unable to divulge. The broken cowrie demonstrated to me a stark reality: the creatures that make their living on this deceptively serene shore confront a world of dangers and challenges, each of which must be successfully overcome through adaptation. Predators of all sorts prowl these pools and sand patches. Powerful typhoons destroy coral heads and hurl massive boulders across the reef crest. For slow-moving animals like snails, searches for scarce food and mates entail a high risk of being eaten, and may in any case end in failure. Ordinary acts expose cowries and a hundred other small animals to ever-vigilant enemies. Somehow, these creatures keep all the risks at bay long enough to grow up and reproduce, ushering a new generation into the world.

But the most enduring lesson that day—one I perceived only dimly at the time—was that the productions of nature, however well adapted they appear to be, are imperfect. Animals fall victim to predators despite well-tested defenses, and predators often fail in their attempts to locate, catch, or subdue their prey. Winds topple trees in the forest despite effective anchorage provided by the trees' deep roots and great weight. Plants fail to set seed because the right pollinators did not find their flowers in time. Raised with the expectation that nature is perfect and in balance when it is left alone, we are disconcerted to discover that imperfection is a necessary reality, a condition that lies at the very foundation of adaptive evolution. Imperfection is not an aberration, or a momentary lapse in the natural order of things; instead, it inevitably arises when success and failure intersect. Competition between living things implies that one side wins while the other loses. It distinguishes between organisms that are well fitted to the conditions at hand and those whose shortcomings restrict them to less favorable conditions or even cause death. The state of adaptation is enforced by continual testing, and no participants manage a perfect score.

When predator and prey meet, they test each other's merits. If predators killed every victim they chose to attack, none of the prey's attributes of size, shape, composition, or behavior would work as an effective defense, and no defensive response could evolve. But if predators sometimes failed in their attacks, surviving victims might owe their success to the effectiveness of their defenses. If predators are common and their attacks are often unsuccessful, these defenses confer a selective advantage on their victims and therefore emerge as evolutionary adaptations. The costs of failure in this predator-prey interaction are, however, strikingly unequal for the two parties. Failure of the predator to secure prey exacts a small penalty, unless the predator is on the edge of starvation or the prey inflicts significant injury on its attacker. The prey's failure, on the other hand, often means injury or death. This inequality—the so-called life-dinner principle—means that prey have less effect on a predator's performance during an attack than the predator has on the defensive performance of its prey. But the predator has its own enemies. Its ability to secure food is therefore influenced more by its enemies than by its victims. From top to bottom in the food chain, failure is the faithful companion to success.

The criteria for distinguishing workable adaptations from failed traits are often consistent, predictable, and unambiguous. For example, a gazelle or zebra must detect a lion early enough, or run away fast enough, or as a last resort kick hard enough to avoid being killed and eaten. The lion must get sufficiently close to its prey without being noticed, and then pounce and deliver the killing bite quickly to prevent the prey from struggling and potentially inflicting injury. It must then consume the meat without delay, so that hyenas or vultures don't steal the lion's hard-won quarry. Similarly, plants have available to them specific, predictable adaptive "choices" for outcompeting their neighbors for light. They can grow fast, tall, or they can shade their competitors by spreading leaves over them. Chemical means and the use of animal partners also work well: trees can prevent plants from growing up around them or on their trunks by shedding toxic bark, as Australian eucalyptus trees do, or they may employ ants to keep the surroundings clear of other plants as African and tropical American acacia trees do. The details of how these adaptations are achieved or which parts of the organism are involved vary, but success and failure can be predicted from well-established principles.

For other adaptations, especially those related to mating and reproduction, performance criteria can seem arbitrary though no less decisive. The complexity of a mockingbird's song, the length and showiness of a peacock's tail, and the virtuosity of an Australian bowerbird's courting display (a structure of branches and ornaments) boost a male's mating success; but complexity of song does nothing for cicadas, crickets, or frogs. In deer it may be the antlers rather than the tail that determine whether a male mates. All these traits are more matters of taste—like human fashions in clothing, hairstyle, makeup, jewelry, and perfume—than like the more predictable antipredatory attributes of shells or the light-harvesting properties of plants. The association between a trait's expression and its effectiveness in attracting mates may arise by chance, much as the association between the meaning of a word and a sequence of sounds or letters does in language, but the association makes all the difference between imperfection—a failure to mate and reproduce—and success. The key, as always, is whether adaptations work.

Inequality and imperfection, then, appear to be universal and necessary accompaniments to life itself. Whenever organisms interact, one party will almost always gain more or lose less than the other as they compete or cooperate. Very rarely will the outcomes be identical for the participants. Although their fortunes may reverse in the long run, with the underdog persisting longer and ultimately gaining the upper hand, the short-term advantage during interactions tends to belong to the party with the greater power and reach.

This principle applies, for example, to the evolutionary relationship between predator and prey. As a rule, predators exert more intense selection on their prey than prey do on their attackers. Not only do they often kill their victims, but they restrict the times and places in which vulnerable prey can be active. Prey species become important as agents of selection on their attackers only if they are dangerous. Poisonous snakes, stinging wasps, biting crabs, and kicking moose can inflict significant injury on a would-be predator, and could therefore influence the predator's behavior. Mobbing—large numbers of relatively innocuous prey ganging up on an attacker, as happens when songbirds mount a group defense against a hawk—may also diminish the evolutionary advantage that predators hold over their prey.

The price of power and influence is often high. Not only must significant force be expended to wield power, as in many predatory attacks, but there is a large material investment as well. In many predatory mammals, birds, and fish, muscles account for up to 30 percent of the body's weight. Competition for status and mates likewise enforces energy allocation to large, conspicuous, and costly displays. For example, the large claw of male fiddler crabs represents a huge investment, accounting for up to one-quarter to one-half of the animal's mass; yet it functions only to impress females and to fight with other males. The female, unencumbered by such an ostentatious device, can use both of her claws for feeding, whereas the male must make do with only one. This example illustrates the principle that an expensive structure or behavior indicates good health and vigor to prospective mates or competitors. The particular form of the display is arbitrary, but the display itself is an expression of real power.

These realities are not limited to the animal world. Disproportionate power and influence in human society are wielded by people who spend vast sums building palaces, buying expensive clothes, throwing lavish parties, driving fancy cars, and generously donating to good causes. These habits amply demonstrate status and advantage, but they do not guarantee success. Like all other adaptations, these social adaptations and their accompanying expenses are good but imperfect ways of dealing with the present and taming or predicting the future.

Even organisms that help each other are subject to inequalities. In such a mutualistic relationship, one party typically exercises more control over the partnership than does the other, and therefore dominates the evolutionary play between them. For example, in the mutualism between legumes—peas, beans, clovers, and their allies—with nitrogen-fixing bacteria on the plants' roots, the plants gain nitrogen, which they cannot obtain directly from the abundant supply in the atmosphere, while the guest bacteria gain carbon and oxygen. The plant hosts evolutionarily dominate the partnership because, as mentioned in chapter 2, they can impose sanctions on their guests by denying oxygen to them.

Underlying inequalities and antagonisms also compel cooperation in a complex mutualistic relationship among East African acacia trees *(Acacia drepanolobium),* herbivorous mammals like elephants and giraffes, and four kinds of ants. Three of the ants feed on nectar produced by the tree in leaf glands. They live in the swollen bases of the tree's thorns, and protect the tree against stem-boring beetles and other insect pests. Only one of these ant species is present on any given tree. When large herbivorous mammals are experimentally removed, the tree devotes fewer resources to producing nectar and hollow thorns. Under these conditions, it grows more slowly and is more infested by damaging insects. Ants of the three species thus suffer, and often turn to tending honeydew-producing insects, which damage the tree. A fourth ant species benefits from the elimination of browsing mammals because it lives in cavities excavated by tree-damaging longhorn beetles. In this system, studied over an eight-year period by Todd Palmer and his associates as part of an international team of scientists, top herbivores—the big mammals—thus impose cooperative arrangements that change or break down when these controlling agents are removed. Every member in this network benefits, but it is likely that the mutualisms could not have evolved without the high-powered mammals.

Inequalities abound at every scale of biological organization. Ecosystems in which plants and plankton fix carbon by means of photosynthesis subsidize ecosystems that run entirely on food sources derived from photosynthesis. Life on the great abyssal plain of the deep sea depends completely on the steady rain of dead organisms and their excrement falling from the sunlit waters above. It is in the top few hundred meters of the ocean where there is sufficient light for phytoplankton—single-celled life-forms capable of harvesting light and taking up dissolved minerals for photosynthesis—to produce most of the food on which all other living things in the ocean rely. Over the course of evolution, this nutritional subsidy has extended to a subsidy of lineages. The sunlit zone has been the source of most deep-sea groups of organisms, whereas the deep sea has contributed only a handful of cave-dwelling and polar species to the shallow-water ecosystems of the ocean. Land life on the desert shores of Peru, southwestern Africa, and northwestern Mexico is subsidized by marine life in the productive waters just offshore, because seabirds feeding on fish ferry food and feces to shore. Elsewhere, life in the sea benefits from nutrients coming in from rivers that drain rich ecosystems on land. In each of these cases, the more productive system subsidizes, and therefore has a disproportionate influence on, the less productive one.

Even the most egalitarian human societies exhibit inequalities in income and status among individuals, among tribes, among institutions, and among nation-states. Insofar as the price of goods and services is determined by adequate information about supply and demand, producers and merchants possess more information about, and have greater control over, commodities than do individual consumers. Through advertising, they can manipulate demand; and by tracking patterns of what goods and services are sold when, where, and to whom, well-organized companies can predict demand and adjust supply accordingly. Companies simply possess and create a better hypothesis about supply and demand than individuals do, and therefore prices are set largely by them. Their information is, of course, far from complete and may even be inaccurate, but the power it bestows nonetheless remains chiefly in the hands of well-organized enterprises. Only when consumers or labor unions themselves become organized into powerful counterweights to business is this economic inequality lessened or reversed. The important point is that inequalities arising from differences in access to information or power are not just manifestations of human nature, but pervade the whole of the living world.

Before Charles Darwin established the theory of evolution by natural selection in 1859, the prevailing view of organisms was that each living thing is a perfect being, designed for its role in life (and often for human benefit) by God. With such an assumption of perfection, it is easy to understand why the idea of adaptive evolution, which logically depends on imperfection and the universal potential for improvement, would have been so radical at that time. It is a cogent reminder of how our expectations and worldviews are shaped by predispositions and assumptions that are often left unspecified and unexamined.

These arguments came alive for me in the summer of 1974, four years after my memorable encounter with the broken cowrie in Guam. At the Smithsonian, I had made a preliminary survey of the vast collection of crabs preserved in alcohol, and identified several species with huge claws that looked as if they were capable of smashing shells. Guam seemed like the right place to look for some of these crabs and to observe how they attacked and broke shells. Lu Eldredge had helped to establish the University of Guam Marine Laboratory, situated just up the hill from the reef flat where the money cowrie was found. With some help from the local masters students, Edith and I obtained several adults of three species soon after our arrival on the island.

Our crabs were magnificent creatures, as alien to our temperate-zone sensibilities as the tropical shells they routinely crushed. One of their two claws—usually the right one—was enormous, studded with hugely thickened molar-like teeth along the length of the pincers. With the victim's shell held firmly between the teeth of the claw's fingers, the crab contracted its powerful claw-closing muscle, the shell shattering with a bang that could be heard from afar. As is often the case with animals whose bite is powerful rather than fast, the crabs were remarkably slow, more like lumbering tanks than light cavalry. As long as we handled them with our fingers firmly pressing the claws to the crab's body, there was no danger of being bitten or scratched.

A fourth crab species, the aptly named *Daldorfia horrida,* was so lethargic that it resembled an inert piece of dead coral. The first time we found one beneath a heavy slab of limestone on the reef flat, we took it back to the lab more as an oddity than as a candidate for shell-crushing. Only when we found shells split in half the next morning in the aquarium did we realize that this piece of coral rubble possessed a huge, powerful crushing claw. In fact, it was this species that Lu had observed years earlier breaking shells in his home aquarium.

Why did these crabs evolve such imposing weapons? The life-dinner principle implied that the prey could not by themselves provide the evolutionary impetus to favor these massive claws. Although I have been fortunate not to have been seriously bitten by one of our crabs, it was very clear that the claws were fearsome weapons belonging to animals capable of vigorous combat. Competitive encounters over shelters or mates, as well as aggression toward the large predatory fish that included crabs in their diet, may well be responsible for evolutionarily fashioning claws into powerful crushing instruments. Selection due to the crabs' enemies therefore produced organs of feeding that provided access to exceptionally well-armored prey, which crabs with weaker claws were unable to tackle.

Crabs are by no means the only animals that have evolved weapons capable of inflicting lethal damage. There is a spectacular diversity of headgear—horns, antlers, battering rams, and in the case of some flies even eyestalks—with which arthropods and vertebrates fight over such defensible resources as burrows, shelters, patchily distributed foods, territories, and above all mates. Horns located on the head or thorax enable dung beetles to engage in combat with other males over females or over sites where females lay eggs. Although many of these have acquired additional display functions, which prevent a confrontation from escalating into violent combat, they have never lost their effectiveness as instruments of injury. Males with larger, more elaborate weapons and more massive bodies typically win contests, either by simply intimidating smaller and less well-endowed rivals or by defeating more closely matched competitors by using force. Injury to, or loss of, weapons is common, and testifies to the high costs of combat and to the role of inequality in competition. Weapons very often enable individuals to feed on better defended prey or to gain access to desirable sources of food, but they evolve under intense selection from competitors rather than from food organisms.

The prey cannot, however, be absolved of all responsibility. Work on young American clawed lobsters has shown that, when these animals are fed a soft diet, the claws developed into relatively slender, weakly biting devices; whereas when the diet consisted of hard-shelled prey, the claws became more robust, capable of delivering a stronger bite force. Whether such an effect also exists in more specialized large-clawed crabs is unknown, but it remains within the realm of possibilities that claw musculature and claw dentition can be molded by diet without genetic specification, much as the muscles and bones of human athletes can be strengthened by rigorous exercise and training. An environment in which most available food is encased in shells might be so predictable that genetic mutations specifying massive, highly muscularized claws could stabilize and amplify the large-clawed state, which would further be favored by selection during combat with the crabs' own predators and sexual competitors.

Despite their strength, these crabs meet with their share of failure. Some of the shells we offered them were simply too large to be crushed; others were too thick, and still others had defenses that prevented the claws from gaining sufficient purchase to deliver a lethal bite. Very often—in 91 percent of trials from one species of box crab—victims survived, albeit with substantial damage to the shell.

For our crabs, these failed attacks were an unavoidable if unlucky part of everyday life, but for me they were a lucky break. They showed me how adaptations are favored and evaluated through tests during specific, identifiable events. Once released by a crab, a shell-damaged snail sets about repairing its shell. After a week or two, the shell is whole again save for a jagged scar where the injury had been inflicted. The shell therefore preserves a record of past trauma. From the victim's perspective, the scar is a sign of success, a demonstration that the shell-maker had offered adequate resistance to its shell-destroying attacker. For me, the scar provides a means to identify which features of the shell contribute to that resistance, in the same way that engineers can pinpoint features of a car that enable passengers to survive an accident. If the shell's scar extends only as far as a rib on the exterior or a thickening on its interior surface, that feature likely helped limit the damage and save the snail's life. Moreover, I could look into the lives of snails long dead. It gives me pause to pick up a scarred fossil shell, hundreds of millions of years old, and to know that some long-extinct predator unsuccessfully attacked a snail in an ancient sea that itself disappeared from Earth's geography ages before dinosaurs roamed the world. By counting scars in a fossil population of snails, I was able to estimate how important shell breakage was as a factor in the evolutionary design of shells. Inert fossils came alive to tell stories about their everyday existence during times profoundly different from our own.

Once I had learned to appreciate the significance of scars, I observed them everywhere. More than half the shells of many of Guam's reef-flat snail species are scarred, and some shells—especially those attacked by box crabs—have three or more repaired injuries. Clearly, breakage was common in the lives of these snails, and their shells reflect its prevalence not just in the present, but over the long haul, over the time span of evolution. Many species have robust shells with crack-stopping surface thickenings, reinforced margins, knobs and spines to make the shell more cumbersome for a predator to handle, openings so narrow that many predators cannot enter the shell, and sometimes—as in cowries—a slippery-smooth surface that makes grasping difficult. These shells have architecturally adapted to a consistent regime of violence, of breaking and entering, of potentially lethal burglary in which the snail's flesh would be torn from its self-built fortress.

Imperfection as a necessary accompaniment to adaptation is hardly confined to tropical shells and their destroyers; it is a universal condition of life. Even such storied predators as the African lion fail more often than they succeed. Some two-thirds of attacks by lions on gazelles and zebras in Serengeti National Park in Tanzania end up in the potential victim getting away unharmed. The basic problem facing both attacker and victim is that success depends on doing each of a host of tasks well. Ideally, a predator should be able to detect and correctly identify all victims, catch those it has detected, and overcome the resistance defenses of all the prey it has caught. A potential prey should remain undetected or, failing that, flee or, failing that, thwart the predator's attempts to kill it. Success in each of these endeavors calls for abilities that often clash. For example, the agility needed for effective capture or for successful escape is usually incompatible with heavy armor. Excelling at one task entails diminished performance in others, unless the animal's energy budget—the power of its engine, in other words—enlarges so that both functions can increase without interfering with each other or with other functions. Within the budget's constraints, the resulting compromise means no task is done as well as it could be were it the only task to be accomplished. Bodies and behaviors are therefore inherently imperfect; yet this imperfection also means that improvements in design are possible. Imperfection permits selection, which in turn contributes to adaptation. The state of perfection is unattainable in the real world of economic competition and conflicting demands.

The principle that all adaptations by one party in an interacting pair are susceptible to counteradaptation by the other applies widely in the human realm as well. From the development of weapons of war to our attempts to conquer diseases, few if any of our inventions—or material adaptations—are unimprovable. History is full of examples of rulers' hubris, claiming that the next generation of weapons will be 100 percent effective and therefore capable of securing a permanent and decisive military edge in conflicts with powerful adversaries. King Philip II of Spain had his invincible Armada in 1588, which was unceremoniously defeated by English and Dutch naval forces and battered by bad weather, to boot; and in the 1980s Ronald Reagan had his missile defense shield, which he promised would fully protect the United States from missile attack by the Soviet Union. These claims propound a dangerously misleading myth of invulnerability and superiority, fueling unattainable expectations of perfection and a risk-free existence. Cowries, crabs, and our own military history demonstrate that such expectations are misplaced. The Soviet Union exploded its first atomic bomb only four years after the United States tested and deployed that weapon in 1945. Cryptography based on the generation of random numbers can be cracked by dedicated code-breakers using powerful computers. Emergent human disease agents—Ebola, HIV, and SARS—at first thwart the most sophisticated medical efforts, but even they are eventually controlled and contained, though never fully disarmed. Nothing is forever invulnerable, either in nature or in the human domain.

We ignore such lessons at our peril. For example, the indiscriminate use of antibiotics to treat common ailments in humans and to fight diseases in domesticated animals has hastened counteradaptation among bacteria, the target organisms of these chemical weapons. We are now engaged in a protracted chemical arms race between adaptable bacteria and pharmaceutical companies designing new antibiotics. Pesticides and genetically engineered hosts with new built-in defenses are also being overused. Pests—mosquitoes and tsetse flies carrying diseases, rusts and smuts infecting crop plants, termites destroying wooden structures, and thousands of herbivorous insects attacking economically important plants—have proven to be highly adaptable in the face of chemicals and introduced enemies.

To slow down such chemical arms races, it is important to avoid using deterrents indiscriminately. If most members of a target population are exposed to a pesticide or antibiotic, and if at least a few individuals survive, adaptation to the new chemical environment is almost certain, because there is a clear signal about which a new hypothesis can be easily formulated. A more limited application of pest-control substances, and the use of more than one deterrent, can raise the adaptive hurdle for the target species because the signals on which an adaptive hypothesis is based are neither consistent nor simple. Individuals vulnerable to chemical weapons will remain in the population longer, and the deterrent's effectiveness erodes less quickly. Countermeasures to the control agents would likely still evolve, but not as readily as in cases where the pest population has no substantial refuge from the chemicals.

Farmers and researchers averse to chemical agents have often turned to organisms for controlling pests, but the same principles apply. The record of newly introduced enemy species to combat a pest outbreak is poor, not only because the enemies often do not limit themselves to the target species, but also because, like all other predators, they fall far short of 100 percent effectiveness in killing the intended victims. The intentional introduction of the predatory land snail *Euglandina rosea* from Florida to tropical Pacific islands in order to eradicate the giant African snail *Achatina fulica,* a major garden pest, met with failure everywhere—not least because the intended target was much too large to be eaten by adult *Euglandina.* The introduced predator did, however, decimate populations of small native land snails.

Of course, situations do arise in which one side in a conflict has an insurmountable advantage over the other. Animals and plants brought by humans from large continents to small oceanic islands have historically driven many native island species to extinction because the competitive, defensive, and reproductive traits of the island species, though adequate for the modest demands of island life, were no match for the invaders, whose traits were shaped under the high performance standards of existence in mainland settings. The losing side simply lacked the resources to mount an effective evolutionary counterattack, because even a substantial improvement in performance would have been insufficient to blunt the powers of the invaders. Similar disparities in technology have characterized human affairs, from the Spanish conquest of the Americas to the European takeover of Australia. What makes such extreme inequalities different from arms races is that the winning party need not invest in adaptive improvements to prevail, whereas in cases where the two sides are more evenly matched, both sides in an adversarial struggle must make outlays in material, capital, and talent.

Imperfection implies the ability to improve but does not automatically lead to a better fit. Incentives for better performance come from our fellow creatures in the struggle for life. If that struggle for existence proceeds at a leisurely pace, selection for more effective means to acquire and defend resources will be weak, and the benefits of improved function simply do not justify the cost. Adaptation is sufficient for the conditions at hand; it need not conform to the best solution that could be devised.

None of this encumbered my thoughts as Lu and I reached the reef's edge. Crawling on all fours, I clung to slippery rock formations while water rushed in and out of deep, steep-walled channels on either side. Crabs and moray eels snapped at my fingers as I probed crevices in search of cowries, turban shells, and top shells. Waves crashing nearby sent up jets of spray over my head. The tide was turning, and it was time to leave. Imperfection rules, but I was not about to become a victim of it.

CHAPTER FOUR

Taming Unpredictability

It is an inescapable reality that life is precarious. Every organism has its enemies, and every life-form is at risk of being done in by an accident or by an unforeseen hazard. Like a house or an automobile or a musical instrument, a living body deteriorates with use to the point where it ceases to work if it is not maintained. Adaptation reduces insecurity and promotes functionality by enabling a living body to learn about, avoid, resist, or limit the damage inflicted on it by predictable hazards; but that same process of adaptation may endanger other organisms. Enemies and the everyday dangers linked to weather and the wear and tear of life are predictable for most organisms because the average life-form will experience them once or more during its lifetime and because they have features in common that the individual is able to discern. But what about the circumstances and events that are so infrequent or so extreme that the average individual never experiences them? Can such unpredictable challenges to security ever be incorporated into an adaptive hypothesis, or must life be forever consigned to wither under their assault? In other words, what are the limits on adapting to a world from which risks can never be fully purged?

These questions not only interest evolutionary biologists like me, but also scholars who hope to apply evolutionary insights to human affairs. Unpredictable threats from new diseases, terrorists, emerging military powers, economic meltdowns, extreme weather, environmental destruction, and climate change confront civilization at every turn. Perhaps evolutionary theory and the long record of life on Earth might reveal insights that political scientists, psychologists, and historians—people who primarily study human behavior and human events—have missed. Over the course of the three-and-a-half billion years that living things have existed continuously on our planet, there were many grave crises—the mass extinctions—during which 20 to perhaps 70 percent of evolutionary lineages disappeared; yet ancestors of all the creatures living on the Earth today survived these catastrophes. How did these organisms survive in the face of calamities spaced tens of millions of years apart for which they lacked specific adaptations? Was their persistence through the bad times a lucky accident, a fluke as unpredictable as the crises themselves, or did their adaptations—their hypotheses of the unpredictable, everyday conditions of life—predispose them to deal with situations they had never previously encountered? Could insights into the unusual turns of life's history be applied to the unpredictable challenges we now face?

Rafe Sagarin certainly thought so. He began his career as a marine biologist, but family connections and a long-standing interest in diplomacy drew him more and more to the policy implications of science, especially evolutionary science. I was one of about a dozen scientists and security experts whom Rafe, then working at the University of California, Los Angeles (UCLA), invited to a series of workshops in 2005 on what he provocatively called "Darwinian homeland security." We convened at the National Center for Ecological Assessment and Synthesis, a government-funded think tank housed on several floors of a nondescript office building in downtown Santa

Barbara. Given the appalling antagonism of the recently reelected President George W. Bush and his administration to science in general and evolution in particular, I was skeptical that anyone in government would listen to, or even hear, what we had to say. It was not at all clear whether we could come up with anything that others had not already discovered. Still, the effort was worth a try; the worst we could do was fail.

Everyone agreed that the threat of terrorism—and indeed of other unpredictable calamities—could never be eliminated. Countermeasures to new weapons and tactics are always possible, giving both the terrorists and those trying to defeat them an incentive to escalate their operations. This message might seem elementary and obvious, but it was an important one to communicate to political leaders and the public as a whole. In a culture of lawsuits, regulations, and an expectation that someone will always be there to help when things go wrong, promises of a secure, risk-free world are readily made and easily accepted.

Once this point was established, we moved on to how knowledge about danger spreads. Dan Blumstein, a behavioral biologist at UCLA with boundless enthusiasm for the Rocky Mountain marmots he studied, pointed out that warnings of the presence of a threat become ineffective if many of them prove to be false alarms. As I noted in chapter 2, inconsistent signals simply become noise that must be ignored. Honest and accurate reporting makes information predictable and allows the receiver to construct an effective response. The implications for officials charged with maintaining security are obvious. A permanent state of alarm and a highly stereotyped response—predictable security inspections at airports, for example—would, Blumstein insisted, make it easy for the enemy to circumvent measures taken and to capitalize on public indifference. Predictability in this case is the enemy of an effective response.

The key to creating, and coping with, rare threats is to understand unpredictability. Pioneers in this endeavor were insurance companies, which for more than two hundred years have taken advantage of the law of large numbers to make highly accurate predictions about life expectancy from the dates of birth and death of a large sample of people. Causes and times of death are individually unpredictable, but when all deaths are taken together, they conform to a consistent statistical distribution, from which life expectancy can be calculated with confidence. Just as in a gas whose molecules behave and collide unpredictably, the behavior of the whole population is far more predictable than that of the constituent parts.

The equivalent of the law of large numbers for a historian or a geologist is what might be called the "law of long-time intervals." If we want to understand the causes, sequences, and outcomes of events that lead to the collapse of civilizations, we must sample the past five thousand years of human history, not just the last few centuries. The likelihood that an interval of one or two hundred years will contain a substantial collapse is small, and even if there were a collapse, it would likely be only one; and with a sample of one, comparisons are impossible, so that predictable information about the process of collapse cannot be extracted. Likewise in the geological record, only a very long interval of time suffices to encompass more than a single mass extinction event. In the past six hundred million years, there have been just six or seven global crises severe enough to eliminate 20 percent or more of existing animal lineages. A hypothesis about truly exceptional conditions, whether it is formulated by scientists or through evolutionary adaptation, is thus contingent upon evidence covering a very long time span. Then, with a sufficiently large sample of lineages or human societies that are confronted with rare hostile conditions, we can ask whether there are characteristics associated with survival, and we can make general predictions about which of several alternative strategies would be most effective when the next unpredicted disaster strikes, as it most certainly will.

Animals I had studied in my days as a graduate student offered one solution to the challenge of hostile conditions. On the upper reaches of most of the world's rocky seashores, small periwinkle snails spend much of their time clinging passively to rocks, often hanging on with nothing more than a dry film of mucus. With the soft parts in the shell safely closed off from the outside by an air-tight door, periwinkles routinely withstand days or weeks of starvation while they are out of the water. In this state, snails are practically inanimate, having for all intents and purposes shut down their metabolic machinery. They survive searing heat, drenching rains, and drying winds. Individuals of some species have been kept a year or more in this inert state, only to revive when placed in seawater. In short, these periwinkles are remarkably well adapted to a predictable life of enforced passivity and chronic, unrelenting stress.

Passive resistance to commonplace inclemency is extremely widespread among organisms. In the snails I studied, the inert state characterizes adults for most of their lives, but in the more interesting and informative cases, a state of suspended animation alternates with periods of intense activity. Most commonly, it is seen in life stages such as eggs, seeds, spores, cysts, and cocoons; but adult dormancy during unfavorable seasons or times of the day is also common. Hummingbirds maintain prodigious appetites and very high activity levels by day, but often pass nights in torpor with a lower body temperature that conserves energy. Trees drop their leaves in icy winters or hot, dry summers. Lungfish and many frogs encase themselves in mud underground when the temporary ponds they inhabit dry up. Bears and many rodents hibernate in dens or burrows in winter, when energy expenditures are kept to a minimum in the absence of opportunities to feed.

Evidence from fossil animals indicates that the ability to shut down metabolic machinery and to enter and remain in a low-energy state may have been key to surviving the great extinction crises of the past. At least some, and quite possibly all, of the land-dwelling vertebrates that successfully endured the end-Permian extinction 251 million years ago lived in burrows and may have had the capacity to aestivate—that is, to survive the hot, dry summer in a state of inactivity. In the plankton communities of the Late Cretaceous ocean, diatoms and dinoflagellates that sank to the seafloor during inert resting stages were less affected by the end-Cretaceous extinctions than planktonic foraminifers that lacked such a dormant stage. Crocodiles, which along with low-energy turtles were among the few large vertebrates to have survived this crisis in freshwater and on land, are well known for their ability to endure long periods of starvation by entering an inert state during the dry season.

It is the ability to shut down metabolism, often associated with encasement in an exoskeleton or burrow, and not simply a low-energy way of life, that confers relative immunity to prolonged periods of hostile conditions. During the end-Permian crisis, for example, two groups of invertebrates that suffered exceptionally great losses were stalked echinoderms—sea lilies and their relatives, which filter particles passively from water currents while attached to objects on the seafloor—and brachiopods, bivalved suspension-feeders that resemble clams but are not closely related to them. Living representatives of both groups lead low-energy lives, but they are unable to seal their organs effectively in a skeleton. Clams, or true bivalves can, on the other hand often hermetically seal themselves in the shell, an ability that enables them to pass intact through the digestive systems of predatory sea stars and fish. The extinction rate of bivalve genera during the end-Permian crisis was only about 40 percent that of brachiopods.

Resistance to extinction in lineages characterized by being able to enter a state of suspended animation was thus a coincidental outcome of adaptation to commonplace circumstances like predation, diurnal fluctuations in temperature, and strong seasonality. The hypotheses of adaptation that enabled organisms to deal with predictable dangers also happened to suffice during times of unprecedented, and therefore unexpected hazards. If such hypotheses were conserved in subsequent generations, the ability to weather extraordinary circumstances similar to those during previous extinction crises would become part of the adaptive makeup of the organism even though hazards of such intensity are far too infrequent to have figured in the formulation of those hypotheses.

When it comes to humans, however, the problem with passivity as a means of coping with unpredicted change is that it necessarily reduces economic activity and power, an outcome consistent with the aims of those who would disrupt society through violence. Inactivity and isolation may promote survival and even a long life span, but they are at odds with maintaining a position of economic dominance. They therefore sacrifice short-term competitive advantages for long-term subordinate persistence, an option that human societies, especially those in superpowers and empires, seem unwilling to embrace.

The Norse settlements in Greenland offer a classic example of passivity in the face of deteriorating circumstances. Although Greenland was settled by Norsemen at a time when the climate was relatively mild and livestock farming was feasible, it became progressively more hostile to European-style farming as the climate cooled. With the advance of sea ice, trade with Norway and Iceland diminished, leaving the Norse communities in Greenland isolated. Archeological evidence shows, however, that the Greenlanders stuck with the increasingly unsustainable farming practices of their forebears, and did not adopt technology that would have allowed them to exploit the abundant marine food sources, especially whales, which for centuries had sustained the native Inuit. The colonies held out for another century of so, but they had disappeared before 1400. Hunkering down works as long as there are sufficient resources available to withstand periods of inclemency, but it fails when the environment remains hostile for a long time or when a better-adapted competitor appears on the scene.

Nevertheless, passive resistance is probably a necessary attribute of all lineages that survive mass extinction and of all human societies that experience a natural or human-caused catastrophe. Some degree of shelter from the outside constitutes the first line of defense, and if it is accompanied by the ability to reduce demand for scare resources temporarily, passive tolerance of siege conditions buys time and prevents outright catastrophic failure. It therefore must be a component of the adaptive hypothesis for any unpredictable circumstance. Organisms and societies that come through crises and that give rise to the next hegemony of powerful economic players likely combine passive strategies—reduced use of resources, isolation and protection from the outside, and inactivity—with active ones.

Additional approaches that are more acceptable to powerful entities and more consistent with the maintenance of economic activity extend the temporal and spatial reach of information-gathering capacity, so that even rather rare or very widely scattered phenomena can be drawn into the domain of predictability. Three overlapping mechanisms can bring this about: increasing an entity's life span, expanding areal coverage through enhanced mobility or more thorough surveillance, and accumulating information through memory devices transmitted among individuals and across generations.

The effects of life span are real but modest. A crisis that looms on average every ten years is effectively unpredictable to an organism with an annual life cycle, but is expected at least once during the lifetime of an individual with a life span of twenty years. With a longer life expectancy, an individual samples its environment on a longer time scale, and therefore is more likely to encounter infrequent situations. The potential for incorporating rare events into an adaptive hypothesis of individuals therefore increases with longer life expectancy.

If longevity allows individuals to predict infrequent phenomena and to make rare events part of their adaptive hypothesis, it may also serve to create unpredictability for enemies. Some long-lived organisms concentrate highly vulnerable life stages during brief periods separated by very long time intervals. Predators with relatively short life spans therefore cannot anticipate the appearance of these defenseless phases.

This appears to be the secret to the survival of eastern North America's periodical cicadas. Every thirteen or seventeen years, depending on the species, vast numbers of slow, defenseless, and noisy adult cicadas emerge from burrows in the ground for four to six weeks to mate and lay eggs. The insects are so oblivious to danger that they can be easily caught by hand. They do not bite, and in fact typically make their characteristic mating calls even when captured. The birds that feast on them live only five years on average, so that the rich harvest of cicadas comes only once every two to four generations, too infrequently for the bonanza of accessible prey to become a predictable part of the birds' environment. The rest of a cicada's life is spent underground, where the insect feeds on tree roots and where it is effectively out of reach of birds.

On coral reefs, infrequent and apparently coordinated mass releases of eggs, sperm, and larvae are known for some corals and seaweeds. Like adult periodical cicadas, these life stages are highly vulnerable to predation, in this case by filter-feeders and fish. The mass releases occur annually, but they are brief enough and separated by a sufficiently long time period that predators are unable to take full advantage of this rich resource. Periodic but infrequent plenty is likely to be common as a means of protecting vulnerable phases of the life cycle. It testifies to the effectiveness of unpredictability as an attainable adaptive response by long-lived organisms.

As these examples illustrate, however, the life spans of most organisms fall far short of the time intervals that separate truly rare events. Some bacteria and plant spores may exist in an inert state for thousands of years or even longer, but larger living things rarely reach the age of five thousand years, the maximum lifespan of bristlecone pines in the western United States and of some Pacific reef corals. Even minor extinction events occur only one to ten times per million years, and the mass extinctions, which disrupt ecosystems on a near-global scale, are separated by five million to fifty million years. Events like these can never be directly incorporated into the adaptive hypothesis of an individual organism.

Even if a long life enables individuals to formulate a hypothesis about rare situations, the evolutionary origins of long life expectancy may be linked to the absence of risk rather than to its presence. It is remarkable that the great age of individuals in nature is associated with risk-free environments and with situations in which threats from enemies have been greatly reduced. Some of the longest-lived clams reside in the deep sea, where a combination of infrequent interactions, low food supply, and perpetual cold enforces a slow pace of life. Despite their small size—five millimeters or smaller in length—some of these clams may live for a hundred years or more. Redwoods and bristlecone pines in the western United States and kauri trees in New Zealand are conifers that live for hundreds to thousands of years, either in temperate maritime climates or in mountainous terrain, where life is uneventful and slow. Among insects, the longest-lived individuals are ant queens, whose twenty-seven years are spent hidden deep inside the colony, where they are tended and protected by expendable workers. Among vertebrates, life span tends to increase with average adult size, but several ways of life are associated with greater than expected longevity. Tree dwellers like sloths, lemurs, and tarsiers, as well as flying birds and bats, have long life expectancies for their size. Compared to ground-dwelling mammals of similar weight and similar rates of oxygen consumption at rest, birds capable of flight live three to four times longer, and produce much lower levels of harmful oxygen radicals, which are linked to aging. Arboreal vertebrates, birds, and bats suffer relatively low rates of mortality during early life stages, meaning that older individuals are exposed to more intense selection than those of most other species.

One reason why a long life span may be an ineffective way of taming unpredictability is that the bodies of living things depreciate with age. The wear and tear on muscles, nerves, and all the other systems that enable the body to function takes its toll, and requires increasing outlays of energy to keep the enterprise going. Sooner or later, as in an aging automobile, the costs of maintenance outweigh the benefits, and the body succumbs to old age. Depreciation of a complex, metabolizing body exposed to frequent hazards seems inevitable, and probably places an upper limit on the life span of organisms, especially of economically active ones. The possibility that single organisms could live long enough to absorb events on timescales of millions of years is thus remote indeed.

Nevertheless, a long life span may lead to the evolution of other, more effective means of mastering unpredictability. This potential does not exist in plants and animals that, like bristlecone pines, live in a safe world not of their own making; but it does exist for those that have created safety through their own prior adaptation. For example, birds and mammals have reduced risks for vulnerable young by building nests, providing nutrition for the developing embryo, and caring for babies after their birth. It is from the ranks of animals like these that large-brained, highly mobile species with keen senses and an exceptional ability to learn have evolved. These long-lived animals—cephalopods like squid and octopuses, birds like parrots and crows, and such mammals as toothed whales and larger primates, including humans—monitor and modify their environment extensively. With the exception of octopuses, they are highly social creatures among which information is gathered from a wide area and travels quickly. Knowledge about new situations comes from many sources and quickly spreads, providing ample opportunity for a suitable response.

In the most effective living systems, it is the combination of central control, exercised by the authority of a brain or government; and diffuse regulation, imparted by flexibly linked and semi-independent parts, that enables the system to sense, communicate, record, and respond to novel situations. If the perceptions are reliable, such a system can effectively formulate and test multiple solutions to problems, and inevitable errors can be prevented from spreading. The relatively diffuse system of government in the United States, where the powers of the central government are divided among the executive, legislative, and judicial branches, and where member states are more or less free to pass laws according to regional needs, is a good example of the type of organization that combines central authority and collective regulation.

Life expectancy in our own species is exceptionally long, having more than doubled since the year 1900 in most countries to the current average of seventy to eighty years. Our long life span has evolved culturally in concert with remarkable mobility and a vast array of methods for acquiring, accumulating, and transmitting information. Knowledge and skills spread from one person to another not just within communities, but among widely separated ones as the result of long-distance migration, trade, and conquest. As semi-independent agents working within societies, individuals play a critical role by speaking, teaching, writing, and learning from others. We may think the information revolution began with the computer age, but human knowledge gathered by individuals and accumulated by society already far surpassed that acquired by any other species when modern human behavior became established in Eurasia beginning some forty-five thousand years ago, or perhaps even earlier in Africa.

Communication among individuals and among groups promotes

the accumulation of knowledge and therefore the capacity to adapt to conditions so unusual that the average individual will not experience them, but it also opens the door to the contagious spread of misinformation and poorly considered actions. This problem arises when individuals cease to be independent actors, when central authority squelches individual expression, or when the impulse to conform to prevailing attitudes overrides a willingness to stand alone. The loss of individuality makes society vulnerable to herdlike behaviors ranging from mob violence and warfare to mass panic and allegiance to dictatorships. There is thus a middle road between completely decentralized behavior, in which uncoordinated individual actions effectively prevent the cultural accumulation of adaptive knowledge, and excessive cohesion, which restricts knowledge and action to a very small cadre of leaders.

The dangers of excessive interdependence are on display during economic crises. With the gradual elimination of trade barriers and the quickening of communication, the increasingly globalized human economy permits local or regional losses of confidence to spread worldwide. This happened during the 1997–1998 Asian financial crisis, and on a much larger scale in the 2008 financial collapse, which originated in the United States with a speculation-driven rise in house prices followed by their sudden collapse. Much earlier collapses, such as the aftermath of the tulip mania in the Netherlands in the 1630s, remained regional. Although the economy of the seventeenth century was already much more international than that of a few centuries earlier, there were still enough barriers in place that internal economic failures did not spread. In effect, different parts of the world economy operated semi-independently to a far greater degree than is the case today.

The obvious, if deeply flawed, solution is to restrict communication. Just as people with contagious diseases are quarantined, measures can be taken to curtail the movements of suspected terrorists, to limit the creation and spread of material that the central authority deems harmful, and to regulate speculation. The problem with such solutions is that the powers given to or taken by central authority are certain to be abused. Restrictions on what can be said, known, or traded run counter to the unencumbered expression and exchange of ideas in a liberally democratic society that is not beholden to particular ideologies. Such exchanges are an essential ingredient of a society's adaptability in the face of novel challenges. Societies may simply have to tolerate and repair damage inflicted by contagious ideologies and herdlike economic behavior.

Human societies confronted with novel dangers often seem incapable of a successful collective response, even if the unprecedented situation is known or predicted. In his book *Collapse,* UCLA's anthropologist, geographer, and ecologist Jared Diamond recounts many instances of human societies ignoring unmistakable warning signs of deterioration. As a result of soil erosion, deforestation, overfishing, and climate change, societies often decline because rulers are too wedded to the status quo that kept them in power, because exploiters—farmers, fishermen, timber companies, and polluters—have no economic incentives to change their ways. I already noted the inability of Norse society in Greenland to adapt to the cooling that affected much of the northern hemisphere after 1250 and that made farming an increasingly untenable proposition in Greenland. The collapse of the cod fishery in Newfoundland during the 1990s is another example of a calamity that was foreseen but not avoided. In this case, as in so many others, the short-term interests of individual fishermen to keep exploiting a diminishing resource clashed with the long-term interests of individuals and Newfoundland society as a whole to harvest it sustainably. Everyone knew that cod stocks were plummeting, but even in democratic societies the political will to confront the problem with the necessary individual sacrifices could not be summoned until it was too late.

Today's societies are likely to do no better. Efforts to reverse the effects of overfishing, which has critically depleted marine food resources around the world, remain regional and halfhearted. Public health, especially the fight against new strains of influenza and the near elimination of polio, has been a bright spot in global coordination, but its effects, and those of increased food production, have been to increase population size and per-capita income, exacerbating the already unsustainable exploitation of the world's resources. Global warming and all its attendant threats to world order have been recognized by some governments, but many developing nations, fossil-fuel producers, and economic interests involved in transportation still either deny the reality of global warming or ignore it by continuing to depend on the burning of coal and oil as the primary sources of energy for the economy. The willingness by those in power to ignore or deny scientific evidence and advice is distressingly common among totalitarian and democratic governments alike. Utopian ideologies, which effectively unite groups in beliefs and actions, appear to win when they compete with evidence-based institutions.

Critics of the science of global warming, overfishing, environmental degradation, and other economically important consequences of human activity point to the uncertainties that are an inevitable part of scientific inference. In their view, the possibility that the science is wrong justifies conservative policies, because the investments necessary to mitigate the predicted disaster would be unnecessary and would hurt the economy. The nature of science as an approximation to empirical truth means that uncertainties can never be eliminated, but critics fail to realize that predictions can underestimate as well as overestimate change. Waiting for perfection is a recipe for failure. As I emphasized in chapter 3, perfection is an illusion, an unattainable goal, and ultimately a poor excuse to do nothing.

Just as medicine is widely acknowledged to be essential for policies on public health, other fields of science should likewise be accorded an equally important role in informing policies on climate change, overexploitation of natural resources, and other problems of global extent. The trial-by-success process that underlies both science and evolutionary adaptation is the most reliable method of acquiring critical information and putting it to use. Ignoring it, or placing it after more parochial ideological and political considerations, is unwise. In my view, every government department should contain a science unit to which all significant matters should be referred as part of the formulation of policy. If a conflict arises between political and scientific considerations, a body with representatives from the two sides should be convened to weigh the arguments. It is, I believe, particularly important to include the science of human behavior in such deliberations, because policies of the past have often ignored or even contradicted what we know about human nature and the incentives that would change people's attitudes. For example, it is better to channel selfish behavior toward the common good than to deny individual desires to own things and to pursue one's own goals. As I noted in chapter 1, this is why the human pursuit of science has been so successful and so relatively free of corruption. Humans are individualistic creatures, yet most of us want to do good in the world. Leaders must make clear how sacrifice, such as paying taxes to underwrite universal health care or paying higher prices to underwrite higher wages for poor workers, benefits both individuals and the society to which they belong. Honest explanation founded on reliable evidence rather than on ideology is the tool of choice in crafting policy and convincing the public of its benefits.

Cost-benefit analysis, a favorite tool of managers and policymakers, in which advantages and disadvantages are expressed in terms of monetary expenses and returns, should not be viewed as the only appropriate means of infusing policy with science. As a means of exchange, money is exactly equivalent to energy. Although money and energy are essential to the functioning of the living systems in which they are used, costs and benefits expressed only in units of money or energy fail to take account of time. But the dimension of time is crucial. Costs and benefit often operate on different time scales: constructing a bridge is very expensive in the short run, but the investment provides long-term benefits; and allowing fishermen to harvest cod unsustainably will give the fishermen a short-term advantage of an income, but it is ultimately disastrous for the fish, for the ecosystem in which the cod live, and for the fishermen's livelihood. To incorporate the long-range consequences of policy into cost-benefit analysis, we must express investments and returns in units of power (energy over time) at each of several relevant time scales. By introducing time into the calculations, we necessarily also factor unpredictability and uncertainty into our predictions. With this broader perspective, borrowed from the annals of the adaptive history of life in a world beset by occasional catastrophes, science should and can bring important considerations to the policy table.

Calamities and setbacks will occur no matter how refined our cost-benefit calculations are and no matter how much history has been incorporated into the adaptive makeup of organisms. Risk of failure, whether it is due to everyday dangers or to unanticipated circumstances, can never be entirely eliminated. Effective adaptation in the face of unpredictability must therefore involve the management of risk by reducing destructive effects of the agents of harm.

The last line of defense against lethal threats is to limit the damage done by them. Damage must be spatially limited and rapidly repaired. Many fast-growing flowering plants, for example, can mount specific chemical defenses when herbivores attack their leaves. The necessary compounds can be rapidly synthesized and transported to the critical areas, and withdrawn when they are no longer needed. Slow-growing plants, by contrast, tend to rely on so-called constitutive chemicals or physical deterrents, which are made and maintained in place regardless of whether herbivores happen to be present. The former type of rapid, flexible response characterizes many lineages of flowering plants with high levels of activity, whereas the more permanent and immobile defenses, which are more expensive to produce, typify more ancient plants with lower rates of growth and photosynthesis. Rapid wound-healing and immune responses, seen in active warm-blooded animals, likewise represent flexible, targeted responses to injury. Modern human societies use this same effective strategy to respond to the destruction wrought by storms, earthquakes, and other disasters.

Most important, however, is the strategy of redundancy. Vital functions must be duplicated and dispersed among similar parts, so that if a function is disabled in one part or in one place, a society or living body will not collapse completely. Redundancy is not universal in nature—disabling one of the two wings of a bat or a bird will compromise the animal's ability to fly—but it is very widespread. When a crab loses a walking leg, there are several others remaining, enabling the crab to move almost as effectively as before. Transportation networks in plants and animals are highly redundant. The system of interconnected veins in the leaves of many plants, and the closed circulatory system characteristic of active cephalopods and vertebrates, preserve alternate routes of transport in case part of the system is disabled.

Economists and entrepreneurs have tended to favor the concentration of vital production in one or a few locations. This strategy takes advantage of abundant, cheap labor and raw materials at favored sites, and discourages production at locations where costs are higher. It also exploits the economies of scale: production costs per piece decline as volume increases. From the perspective of limiting damage, however, concentration imperils a system. It is, for example, irresponsible to concentrate the production of essential crops to just one or two regions, for if the crop should fail because of some unanticipated calamity, the destructive effects could cripple the food supply worldwide. Military bases, fuel production, intelligence-gathering, electricity generation, and a host of other key industries should likewise be dispersed even at the cost of reduced efficiency. Excessive concentration is thus both shortsighted and risky.

Indeed, efficiency has not been an overriding standard of performance in active systems, either in living nature or in the human domain. Dominant elements gain a competitive edge by acquiring and expending resources rapidly, an activity that is incompatible with conservation and efficiency. For example, deciduous trees in seasonal climates throw away vast numbers of leaves, which are replaced as conditions more favorable to photosynthesis return. Although some of the nutrients in the old leaves are withdrawn before the leaves fall, there is nevertheless a considerable waste of carbon. This inefficient use of material resources is acceptable because trees can count on ready access to new sources of food during the next season or generation, and because the competitive advantages outweigh the costs imposed by inefficiency. Leaf life spans are considerably longer—up to four years in extreme cases—under conditions in which the cost of replacement is invariably high either because resources are chronically scarce or because low temperatures prevent their rapid uptake.

Warm-blooded mammals and birds release vast amounts of unusable excess heat. This inefficiency is acceptable because these animals gain many advantages from their high metabolic rates. There is little incentive to reduce costs because these animals already enjoy favorable competitive positions. Likewise, the human economy, with its dependence on rapid and cheap transport and power supplied by fossil and nuclear fuels, is energetically very inefficient, but the short-term economic benefits have been enormous. As individuals, American drivers—and probably most other people, if they had the resources—prefer large, heavy gas-powered vehicles whose fuel use is markedly inefficient, because owning and operating such vehicles is seen as indicating wealth and status. In evolution as in human affairs, efficiency matters to the competitively disadvantaged, but not as much to winners. In many situations, including the maintenance of redundancy in an unpredictable world, efficiency often takes a backseat.

But does any of this matter in the long run? What if the persistence of life on Earth, or of human civilizations, is purely a matter of luck, of living in a place where the predictable agencies of destruction have not reached? We all know of a thousand anecdotes in which individuals survived a hideous calamity because people had missed the plane that crashed, ducked just in time to avoid being struck by a bullet, were out in the ocean on a boat while a tsunami leveled their village, or were imprisoned in a dungeon insulated from aerial bombardment of a massive volcanic eruption. The role of chance is undeniable at every scale of severity of hazards, with the result that fate is not rigorously determined by the presence or absence of adaptations. There is a random component to survival even in the presence of effective adaptations.

Ironically, it is impossible to gauge the role of chance in surviving catastrophes if we cannot identify the causes and locations of those catastrophes. Without such an understanding, chance becomes an explanation of last resort, in effect an expression of ignorance and inadequacy. We therefore need a theory—or at least a well-founded hypothesis—of catastrophe, an explanation that encompasses disasters in the history of life as well as major disruptions in human civilization. The analysis of past extinctions and of the demise of vanished civilizations forms the basis of just such a theory.

Few topics have been as hotly debated in paleontology as the causes of the great mass extinctions. Disagreements have centered on whether the underlying cause is gradual or sudden, to what extent the various collapses are comparable in cause, whether the magnitude of loss is correlated with the magnitude of the trigger, or how such events as extraterrestrial collisions, major volcanic eruptions, fluctuations in sea level, and climate affect vulnerability to collapse. Amid all the uncertainties, however, the following conclusions, with direct bearing on the role of chance in catastrophes, appear to be well supported.

Great disasters—mass extinctions and civilization collapses—are associated with highly unusual, life-unfriendly conditions. These conditions—massive volcanic eruptions, collisions of Earth with celestial objects, and in the human case crop failures brought on by drought or excessive rainfall—interfere with photosynthesis on a large scale, and therefore jeopardize all those consumers that depend on this process. Feedback loops between producers and consumers (which we will examine further in chapter 7) magnify the effects of these conditions because their dissolution brings about collateral losses of species that are not directly involved in the feedbacks. Adverse conditions become disastrous only if their intensity crosses some threshold—a tipping point of sorts—beyond which damage spreads uncontrollably. For example, an ordinary rise in the concentration of carbon dioxide need not be catastrophic, but if it accompanies or is a consequence of other problems—such as the upheavals of a huge eruption, or widespread human-caused environmental destruction—it may contribute to an overall economic collapse.

If the great crises of the past were triggered by events disrupting food production, then species that survived would either have lived in rich habitats unaffected by the events, or have already been so chronically starved that a photosynthetic disaster had little effect on them. Either way, survivors had to live in refuges. Few if any productive ecosystems would have been untouched by the global events that brought on mass extinctions; but chronically food-poor systems were likely widespread, as they are today. In such environments—caves, the deep sea, and some polar habitats—most species grow slowly and leave few offspring. In the aftermath of a crisis, such species are poorly situated to expand rapidly, especially when compared to species that weathered the economic downturn by entering into a state of prolonged inactivity and recovering as active players when more favorable conditions return. In other words, chronically stressed species are less likely to compete effectively for resources when photosynthesis resumes on a large scale. Only those survivors whose life cycle alternates between an inanimate state in a resistant seed or egg and a fast-growing, fecund stage of high activity are poised to take full advantage of favorable conditions. Species no doubt sometimes survive through times of extinction by the luck of living in unaffected places, but it would be a stretch to conclude that luck is the preeminent determinant of success. Surviving an extinction event is one thing; contributing to subsequent recovery is quite another, requiring rapid individual and population growth, dispersal out of the refuges, and the ability to capitalize on newly available opportunities. If a population is reduced to just a few individuals as the result of a catastrophe, it will remain vulnerable to extinction, because even the most minor setbacks could wipe out all survivors. Long-term success thus depends on recouping population losses rapidly.

If chance were the main arbiter of surviving rare challenges, there would be no cumulative adaptation to such events, and the magnitude of extinction would fluctuate from event to event without displaying a discernible trend. No reliable data are available for species-level extinction, but most statistical analyses that have been carried out point to a general decline in the magnitude of extinction during successive mass extinctions as well as over the last five hundred million years taken together. The celebrated end-Cretaceous disaster, which spelled the end of the dinosaurs, affected fewer lineages (31 percent) than such earlier mass extinctions as the end-Triassic (39 percent) and end-Permian (70 to 80 percent). If such a decrease proves to be real, it could mean either that more lineages live in locations that are safe from the usual causes of mass extinction, or that more lineages have over time incorporated traits that protect them against such extremes.

What, then, does evolutionary biology have to say about Darwinian homeland security? For me, the conclusions that emerged from our coffee-drenched deliberations in Santa Barbara capture all the essentials of adaptation. The first requirement is that we accurately identify and monitor potential threats. Passive resistance must be part of any response, but for our economically active species, more energy-intensive responses—aggression, intervention, and so on—are the preferred solution. Most important, however, are principles of organization—flexible and diffuse control, redundancy, and the ability to restrict damage—as well as analytical methods that treat policy options at short and long time scales. In short, the most effective adaptation for dealing with unforeseen circumstances is adaptability.

Insecurity remains a problem for all life. Unpredictable situations will continue to arise, either as unforeseen physical catastrophes or as adaptive measures by living things seeking an advantage over rivals. Uncertainty and insecurity will not go away, but adaptation can reduce their frequency and ameliorate their effects. Evolutionary biologists may not uncover the magic bullet that policy-makers grappling with human-caused insecurity are looking for, but they—and by extension all the rest of us—have access to three-and-a-half billion years of collective experience with hazards of every kind. We may not wish to copy nature's solutions slavishly, but we would be foolish not to consult the rich diversity of adaptations that organisms have tried and tested in nature's realm.

No adaptation can persist, and no information can accumulate, without some form of transmission, or inheritance. Improvement and modifications cannot occur if an adaptation, or a scientific hypothesis, dies with the death of its bearer. How inheritance is achieved is the subject of the next chapter.

CHAPTER FIVE

The Evolution of Order

For as long as I can remember, I have been drawn to the almost mathematical precision of form that is found in nature. The pattern of veins on the underside of a leaf, the five-part symmetry of a flower or a starfish, the spiral arrangement of scales in a pinecone, and the remarkably even spacing of riblets and tubercles on spiral seashells can all be described with simple equations, presumably because living things develop and grow according to a few basic rules. What are these rules, and where do they come from? What roles do genes and the environment play in creating the order that is so strikingly on display in the structure of living things?

The order found in nature can be so seductive that signs of disorder are easy to ignore or dismiss. Yet it is the rule-breaking departures from expected regularity that hold the key to how adaptation begins. I learned to take exceptions seriously when Adolf Seilacher, an immensely gifted and unconventional German paleontologist well known for his cigar-smoking habit, came to give a talk at Princeton in 1967, when I was an undergraduate. Seilacher was a master forensic scientist, extracting clues about the lives of fossil animals from preserved skeletal remains. He was particularly interested in how the growth and form of an animal are influenced by the physical forces that an animal encounters as it goes about its daily routines, and by the forces that the animal applies with its own muscles. Some of his work dealt with ammonites, fossil shell-bearing animals distantly related to living squid and octopuses. The symmetrical shells of ammonites indicate a swimming habit, in which the plane of symmetry is vertical. By expelling water forcefully from a funnel-like opening, the living animal swam through the water using the principle of jet propulsion. Some aberrant individuals had become asymmetrical late in life because an organism had settled on the ammonite's right flank. The extra weight caused the animal to tilt to its right side. To compensate for this asymmetrical load, the ammonite began to add new shell material somewhat to the left of the plane of symmetry, enabling it once again to hold the shell in the normal vertical position. Normal symmetry in this ammonite was therefore the result of the symmetrical distribution of forces on the right and left sides of the animal as it swam. It was thus the forces acting on a moving animal, and not only the underlying genetic instructions in the animal's DNA, that determined form; and it was the exceptional and accidental growth of foreign organisms on the shell's exterior that provided the evidence.

Once I was primed to think about exceptions to the apparent orderliness in nature, I found it everywhere. For example, almost no organism is perfectly symmetrical. Our bodies have approximately equal right and left sides, but most of us show a distinct preference for one hand (usually the right one) over the other, and internally our anatomy is strikingly asymmetrical, with the heart lying toward our left side and our intestines coiled in an irregular configuration. Seashells, which are basically hollow tubes that expand from the narrow closed end to the broad open end, are coiled in a spiral that closely matches the so-called logarithmic spiral. This spiral form results whenever a structure grows at one end (the opening) without changing shape. But almost all shells depart from this ideal form, changing shape slightly or greatly during the course of spiral growth. There may be rules of growth and form, but they are not absolute. There is clearly variation. Some of this variation must be under the control of genes, but some is little more than an expression of developmental noise, an indication that growth and form are not strictly regulated.

There exists in the realm of living nature a palpable tension between orchestrated order and unregulated variation. On the one side are well-established adaptations that are finely tuned and controlled by the body's metabolic machinery, which is ultimately specified and regulated by the expression of genes. Development of the body from a fertilized egg through a series of life stages to the adult proceeds in a highly ordered sequence in a physiological environment that is unlike the environment outside the body. The process resembles the assembly of an automobile under the precisely controlled conditions of an assembly line on the factory floor, except that every step in manufacture was conceived and monitored by sentient engineers and not by the feedback between genes and a living body.

On the other side are variation and flexibility. No two organisms are exactly alike, even if they are identical twins with exactly the same complement of genes. Some of this variation is therefore due to environmental influences; it is basically nongenetic noise, reflecting the absence of precise specification and regulation. Other variations arise from mutations, or slightly altered genes. It is these genetic variations that constitute the raw material on which natural selection can act. But there is still a third category of variation, the kind that is the result of adaptation itself, a derived versatility in a genetic system that promotes quick, flexible responses. This is the stuff of learning, creativity, and innovation.

If the rules and regulations governing the genetic system and the developing embryo are too strict and too inflexible, there can be no variation and therefore no gene-based selection or adaptation.

Without rules, there is just chaos. Selection and adaptation must operate somewhere in the middle, in a system where there is order as well as both genetic and nongenetic variation. In order for evolution to take place, modifications must be passed on through the generations; they must be heritable in some way. But we also know that form and physiology are influenced directly by forces and by the chemistry of the environment. For effective adaptation to occur, these environmental influences must be incorporated and controlled by genes, the molecular vehicles by which information is inherited by organisms. Gene-based adaptation evolves when genetic control—the equivalent of the laws and norms that govern civil society—is established.

The evolution of order from noise has interesting parallels in the human realm. From the development of economic markets and legal systems to the evolution of languages, the emergence of rules and regulations seems to take the same course as in the evolution of organic form. To understand the origins of adaptation, we must probe how rules arise, and how a highly regulated system such as the genome or our system of laws yields adaptive flexibility, creativity, and the ability to learn. This is the task to which I set myself in this chapter.

My entry into this line of inquiry was, to say the least, indirect and unanticipated. It all began when I was teaching a graduate course in marine biology at the University of Maryland early in 1980. Sitting in my cramped office, I contemplated what I should talk about in my next lecture. I had promised to discuss introduced species, plants and animals that people brought to parts of the world not previously inhabited by those species. Rather than take the more usual tack of summarizing how new immigrants affected the ecosystems they invaded, I thought it would be interesting to ask whether and how immigrants adapt to their new surroundings, and whether and how native species evolve to cope with the new arrivals. These questions had rarely been asked then, so the lecture would have to be an exercise in speculation. This did not trouble me in the least, however, for

I have always believed that courses should introduce students to new ideas and new questions rather than merely review information that was readily available elsewhere. Besides, I thought I might become so excited by the material that I might be tempted to conduct some original work. And that is exactly what happened. I would start with a simple question about what I suspected would be genetic adaptation of an introduced species, but I would end up thinking about the much more intriguing question of how adaptations arise from unsupervised variation and turn into a well-regulated genetic modification.

I settled on the common periwinkle *(Littorina littorea),* a homely snail that is perhaps the most abundant animal on contemporary shores of southeastern Canada and New England. This European species was entirely absent in North America before British ships brought it in ballast from northern Ireland and Scotland to Nova Scotia shortly before the year 1840. Once established there, the species quickly spread north to Newfoundland and south to Delaware. By the late nineteenth century, the common periwinkle was an important consumer of seaweeds, a major food item for predators (crabs, fish, and birds), and a prolific source of empty shells for native hermit crabs. Despite its prominent place on American rocky shores and salt marshes, the periwinkle seems not to have displaced any native species or driven any species to extinction. Did the periwinkle undergo detectable adaptive evolution upon its arrival in North America, and how did it evolutionarily affect the other species with which it interacted as competitor, food item, or consumer? In the lecture, I suggested that some genetic differences between European and American periwinkles could have emerged, and that some of these might reflect adaptations of the American population to the novel conditions there.

This case was especially intriguing in the light of another European immigrant to Atlantic North America, the green crab *Carcinus maenas.*

This crab is a voracious shell-breaking and shell-entering predator of small snails, including young common periwinkles. The green crab had been present on American shores between Cape Cod and New Jersey as early as 1820, but it was not found north of Cape Cod until about 1905. During the first half of the twentieth century, the species spread northward, reaching the Atlantic coast of Nova Scotia by 1954. Late in the century, it had extended as far as the Gulf of St. Lawrence, but it remains absent from Newfoundland. Snails surviving a crab's shell-breaking attack are able to repair the damage inflicted on the shell, as indicated by the presence of jagged scars visible on the exterior. Green crabs arrived in Maine and the Maritime Provinces fifty to one hundred years after the common periwinkle became established there. If crabs often attacked victims they were unable to kill, their arrival on these shores should coincide with an increase in the frequency of scars in populations of the periwinkle. In Newfoundland, where the green crab has so far not been found, the frequency of repair should have remained unchanged over the same period. Moreover, if green crabs were important new selective agents, thin-shelled periwinkles should have become less common upon the arrival and establishment of these crabs. To test these predictions, I would need old, well-documented samples of periwinkles as well as more recently collected material, together with dates of arrival of the crab at various locations north of Cape Cod.

Fortunately, a wealth of suitable material existed in major museums thanks to the foresight of early naturalists, who believed in the value of collecting even the commonest and least eye-catching species. There were hundreds of samples of the periwinkle, all provided with locality information and the date of collection, from throughout the range of the species, some dating back to the middle of the nineteenth century. I, too, had made substantial collections of the common periwinkle on my visits to shores from New Jersey to New Brunswick.

At least some of my predictions were confirmed. After examining thousands of specimens, I found that samples collected between Cape Cod and Nova Scotia before the arrival of the green crab contained an average of .053 scars per shell, whereas samples from the same stretch of coastline taken after the crab had become established had an average of .106 scars per shell. There was, in other words, a doubling in the incidence of unsuccessful predatory attacks on the common periwinkle after the green crab had begun to spread northward. Also as predicted, there was no change in the very low incidence of repair in Newfoundland, where green crabs have not yet become established.

Consistency of an observed pattern with prediction does not, however, mean that the proposed underlying explanation is correct. Given that green crabs and periwinkles have coexisted on European shores ever since the periwinkle first arrived there some two-and-a-half million years ago from the North Pacific, European populations of periwinkles should not have witnessed the increased frequency of scarring that I had detected in American populations north of Cape Cod. The American museums whose collections I had used for establishing the pattern in populations from New England and adjacent Canada lacked sufficient samples from Europe. What a great excuse to visit my family in the Netherlands. My wife and I even invited my parents to vacation with us on Schiermonnikoog, a lovely, quiet island of dunes, wide sandy beaches, and mudflats bordered by dikes, located in a line of barrier islands separating the Wadden Sea from the North Sea. The dikes were perfect habitats for the periwinkle.

Thousands more periwinkles passed through my fingers. Again to my delight, the results mainly accorded with my prediction. There had been no changes in the frequency of repaired shells in Norway, the Netherlands, and France. Only in inland Danish waters was a change detectable: the frequency of scarring had inexplicably declined there during the twentieth century. The hypothesis that periwinkles experienced more unsuccessful predation after the establishment of the green crab in North America north of Cape Cod remained viable, though of course other contributing factors—such as changes in the abundance of other shell-damaging predators—might still be important.

Green crabs should also have affected native species, those with a long prehuman history in North America. One such candidate was the dog whelk, *Nucella lapillus.* Examination of museum samples revealed that the frequency of repair in this species had increased at each of six sites in North America from which specimens collected before and after the green crab's arrival were available. Moreover, unlike the situation in the common periwinkle, in which I observed no obvious change in shell thickness over time, the dog whelk samples displayed a small but consistent increase in shell thickness, an adaptive response that was to be expected if shell breakage became a more important agency of selection. Subsequent work by Jonathan Fisher and his colleagues has shown that there was also an increase in shell size. In still another native snail, the flat periwinkle *(Littorina obtusata),* usually found crawling and feeding on brown seaweeds, shells with a broad opening and thin wall were replaced at sites with green crabs by thicker shells with a narrower opening. Of the species studied so far, the flat periwinkle was the most dramatically affected by the green crab's arrival. As the smallest species, it is probably the most vulnerable to predation by the green crab.

When I began this project, I expected to find adaptive responses to the establishment of green crabs in North America. And indeed, appropriate changes in shell form did emerge in at least two species of snail over a time interval of less than fifty years. Along with other evolutionary biologists, I more or less automatically equated adaptation with genetic change. We assumed that shell form was genetically inherited, and that either a favorable new mutation had arisen or selection by the crab had eliminated thin-shelled genetic variants. But was this really the case? There was evidence of change, but no evidence for genetic change.

Had the response indeed been genetic, the history of New England's shore snails and their predators would have gone down as one more nice example of natural selection. But, as so often happens in science, the situation proved to be far more interesting than our expectations. Four years before I began my work with periwinkles and dog whelks, Edith and I spent several months at the Smithsonian Tropical Research Institute in Panama. There, we met A. Richard Palmer, a brilliant graduate student at the University of Washington who had come to Panama to study predation of snails by fish. He had begun to suspect that snails were able to respond to predators by altering the pattern of shell formation when they detected enemy odors. At the time, I dismissed this idea more or less out of hand because it conflicted with the uncritically accepted view that genes are responsible for most evolutionary change. In the late 1980s, however, Rich began some very careful experiments with dog whelks, in which he showed that shells became thicker and developed a narrower opening in the presence of crab odors, but without the actual crabs being present, relative to control snails that were not subjected to the scent of crabs. Experiments by others showed that periwinkles also responded architecturally in the direction of greater resistance to predation when they were exposed to chemical cues from crabs. The mere presence of chemicals indicating danger from predators thus caused a subtle change in shell growth. Apparently, snails sensing danger spend less time moving, feeding, and growing, with the result that new shell material was added to the interior rather than the edge of the shell. The shell thus grew more slowly, and the direction of growth changed in such a way that the shell opening expanded less quickly as the animal grew than was the case in snails that perceived no risk of being eaten by crabs. All these effects were nongenetic, exemplifying the phenomenon that biologists call phenotypic plasticity, the sensitivity of gene-controlled form to environmental conditions.

Although these responses are stimulated specifically by chemicals of animals that the snails recognize as dangerous predators, the response itself is a very general one, occurring whenever growth rate decreases. In mussels and other bivalves, for example, slow growth caused by old age, prolonged inactivity, scarce food, or all the animal's resources being diverted to reproduction always results in an inflated shell; whereas fast-growing individuals have a flattened shell. The generality of this relationship indicates that shell form is closely tied to growth rate. It is an example of variation imposed by the factors that control growth. Many shell-builders surely capitalized on this variation by exaggerating differences in growth direction according to whether shell growth was slow or fast, but taking adaptive advantage of this relationship means careful regulation of growth rather than the passive dependence of growth on the animal's condition. In many snails, for example, the ancestral pattern of continuous but slowing growth with age has been replaced either by episodic growth, in which periods of rapid increase in size alternate with intervals of no growth; or by so-called determinate growth, in which the animal ceases to grow entirely when it has reached reproductive maturity. Once shell growth in the spiral direction has ceased, the animal constructs all sorts of shell elaborations, such as obstructions of the opening, spines, and a winglike expansion of the shell's outer margin. The key point is that these more precisely specified and controlled variations, reflecting adaptive modifications on the theme of ancestral spiral shell architecture, could not have arisen, or been favored by selection, without the existence of unregulated variation resulting from the direct effects of the environment on growth and form.

The idea that apparently adaptive changes arise from nongenetic variations was hardly new, but it did clash with the strongly gene-centered view of adaptation that biologists have held from the 1930s to the present day. In the early twentieth century, the polymath Scottish scientist and classicist D'Arcy Wentworth Thompson gathered dozens of examples of biological structures whose form depended directly on the forces to which animals and plants were exposed or which those organisms themselves exerted. Many more cases have come to light since. The dandelions that used to grow in my lawn had the rosette of basal leaves splayed out over the ground. This growth form protects the plants not only from a heavy-footed primate, but also from the lawn mower. Elsewhere, dandelion leaves rise vertically from the ground. The difference seems to have nothing to do with genetics and everything to do with forces to which the plants are exposed during growth.

In our circulatory system, genetically specified proteins regulate the formation of capillaries (the smallest blood vessels), but the placement of those capillaries is dictated by the body's internal environment. Capillaries form only where oxygen levels are low. Moreover, the size of blood vessels is determined by the strength of local blood flow. Arteries—the largest vessels—form where flow is strongest, capillaries where it is slowest. Bone growth and repair take place in directions dictated by the animal's own movements. This is why exercise is so important for healing injuries to limb bones.

Thompson was so impressed with the power of physical forces to explain the intricacies of form in living things that he considered natural selection as superfluous, and he rejected evolution itself. This was clearly an overreaction stemming from Thompson's underestimation of the power of natural selection to modify the relationships between living bodies and the environment. Adaptations may exaggerate the role force plays in controlling the growth of bone, but they also have the power to counteract or neutralize that role. Under natural selection, few if any relationships between physics and organic form are absolute and unchangeable. An organism's metabolism and activity, which are controlled and regulated to a large extent by genes and which are therefore subject to adaptation, affect the evolutionary "conversation" that living things carry on with their surroundings.

Despite the clear effects of the environment on the expression of form, the basic body plan is certainly under genetic control. Spiral shell growth, leaf development in plants, body organization in arthropods, and the pattern of formation of bone and other tissues in vertebrates are among the innumerable examples of traits and processes strongly determined and regulated by genes. The ability of animals to detect and respond to cues issuing from food or predators is likewise orchestrated genetically. Without that ability, snails exposed to odors emanating from dangerous predators might be unable to react by altering growth and form in their shells. In other words, the flexibility of response seen in our snails and in countless other organisms is built on a genetic infrastructure.

But the traits we observe in many species are controlled not by the genes of those species, but by the sinister machinations of parasites. For example, gall wasps take over the form-constructing machinery in host-plant leaves to produce the hollow structures in which their larvae develop. Parasitic worms whose complex life cycles involve two or more different host species face the problem of being transferred from one host to another. Many have solved it by disabling or altering one host so that it becomes conspicuous and therefore more likely to be eaten by the next host. Here again, the traits of one species are hijacked by another species that manipulates it for its own ends. Such genetic takeovers may well be the rule in situations in which one species acts as host and the other as either wanted or unwanted guest.

One intriguing possibility, hinted at by several lines of evidence, is that adaptive novelties pass through an early evolutionary stage in which their expression is provisional and largely under environmental control, with genetic regulation playing little or no role. As their bearers come to depend more and more on these new traits, regulation shifts increasingly to the complex machinery of chemical regulation orchestrated by genes. Initial tentative and noisy expression gives way to a more orderly, more precisely specified and thus more predictable state of novelty that is well integrated into the overall physiology and architecture of the body.

Fascinating examples of this increased rule-making come from the transition of symmetrical animals, in which the right and left sides are mirror images, to asymmetrical descendants, in which one or the other side becomes larger and takes on a side-specific function. In the early evolutionary stages of asymmetry, the side that becomes larger may be either on the right or the left and is not specified; larger-left and larger-right individuals occur with equal frequency. Clawed lobsters exemplify this stage. The larger claw is stouter, and has the opposing surfaces of the closing fingers studded with larger projections, than does the smaller claw. Food is held in the small claw but generally dismembered by the other, larger claw. In the American lobster, the larger claw appears on the right side about half the time. In most true crabs and in several groups of hermit crabs, the direction of asymmetry has become consistent, with the larger claw normally developing on the animal's right side. Only when that claw is lost during a fight will the claw on the left become modified in successive molts to take over the lost right claw's function while a new small claw regenerates on the right side. The consistent handedness of these crustaceans appears to be genetically determined, unlike the situation in the American lobster. A. Richard Palmer has shown that in these and hundreds of similar cases of transition from unspecified to specified handedness in animals, there is a shift in timing in the expression of asymmetry. Unspecified asymmetry appears in late stages of an animal's development, whereas genetically specified, consistent asymmetry appears early on.

A similar sequence of events probably took place in the evolution of snails. The shells of most snails are consistently right-handed, with the shell's opening appearing on the observer's right when the shell is held with the coiled portion up and the opening toward the observer. Shell asymmetry appeared during the Early Cambrian period, but during that early evolutionary phase the direction of coiling was apparently not fixed, with both left- and right-handed shells appearing in comparable numbers. In these early snail-like shells, asymmetry probably appeared at a late stage of development. As the direction of coiling became increasingly consistent, its expression moved to earlier larval stages. In living snails, mutations leading to a reversal from right-handed to left-handed are already expressed in the earliest stages of development, and may in fact be encoded in the egg of the mother snail.

The behavior of individual animals may hasten the transformation of a trait from nongenetic control to genetic specification. For example, animals observing other members of the same species mating will often mimic the sequence they see, including courtship displays. These learned routines are then passed on to the next generation and thus become a fixture for members of the lineage, setting the lineage apart from others in which the routine may be slightly different. The routines may become so different that members of the two lineages would not recognize, or mate with, each other. This kind of isolation can lead to the formation of genetically separate species. A similar mechanism likely accounts for the origin and divergence of dialects in birdsongs and human languages. A local variation in song or speech in an isolated population is learned by members of the group and is culturally inherited. In time, the local variant diverges from the song or speech in the parent population to the point of incomprehension, and any genetic contact that might have existed between these groups is broken.

Paleontologists have long recognized an increase in the regulation of form over the course of the evolutionary history of particular branches in the history of life. When my Davis colleague James A. Doyle and Yale's Leo G. Hickey began to study leaves of some of the earliest flowering plants, which grew along the banks of what is now the Potomac River in Maryland and Virginia some 110 million years ago during the mid-Cretaceous period, they noticed that the tiny veins connecting the leaf's major veins varied greatly from leaf to leaf. In these early leaves, there was no fixed pattern of where and how many veinlets would form to fill the network of the larger conduits. Lineages derived from these Cretaceous ancestors achieved much greater consistency in the position and expression of the leaf's minor vascular elements, indicating that the process of leaf and vein development was now much more controlled than in earlier representatives.

There is a similar story for trilobites, a group of arthropods that occupied marine habitats throughout the Paleozoic era, from 542 to 251 million years ago. In Early Cambrian representatives, living at a time of enormous ecological expansion, individual trilobites of any given species varied greatly in the number of body segments and in other characters of shape. Later species were much less variable, presumably because many aspects of form, including the number of segments, came to be more precisely specified through genes acting during the course of the animal's development from egg to adult.

The shift from a state of unregulated development to a genetically more regimented cascade of constructional steps is not understood well in detail, but biologists have given it a name: genetic assimilation. This phenomenon raises three important questions. First, are there environments that favor—or allow—unregulated construction to flourish? Second, under which conditions is genetic assimilation favored? And finally, how is adaptive flexibility in behavior and form achieved in a highly regulated, rule-bound system?

To answer the first question, building on ideas first proposed by the late Harvard evolutionist Stephen Jay Gould, I suggest that the initial evolutionary stage of unregulated innovation occurs in organisms with a weedy life history, in which a resistant seed, egg, or spore alternates with a short-lived, fast-growing phase. The ephemeral body has a flimsy, imprecise construction suitable for taking advantage of briefly favorable conditions but not adapted for sustained defense. The earliest members of many major branches on the tree of life had exactly this construction: they were small, short-lived creatures that achieved reproductive maturity at a stage that would have looked juvenile in an ancestor. In such a body, the rigidities that accumulated in the ancestor were loosened, meaning that innovations first expressed as direct responses to the environment do not interfere with the few remaining genetic controls on body form inherited from the ancestor. Early flowering plants, which evolved about 140 million years ago during the Early Cretaceous period, are thought to have been stream-side weeds with a truncated life history, enabling them to occupy competitively subordinate positions in habitats where water and nutrients were plentiful. Other examples of major groups with early small, fast-growing ancestors include moss animals (bryozoans)—colonial filter-feeders often found on seaweeds, stones, and shells—as well as molluscs, dinosaurs, and mammals.

The constructional freedom and flexibility implicit in the throwaway body of a fecund, fast-growing organism confer phenotypic plasticity—sensitivity of form to environmental influences—because of a lack of regulation. Lax control works only in situations where resources are abundant, competition is weak, and the risk of being eaten is low. The cold waters of polar seas satisfy these criteria, and it is no accident that most marine species in which phenotypic plasticity has been documented are from cold-temperate, Arctic, and Ant-arctic coasts, and not from the tropics, where threats from competitors and predators are chronic and severe. Shells of cold-water clams and snails are notoriously variable and often sloppily constructed, in great contrast to the precision of form in tropical shells. The architectural freedom allowed in the cold and in other biologically safe and resource-rich settings would seem to be favorable to evolutionary innovation, but new adaptive states will not be realized in the absence of intense selection by enemies. Without the incentive of competition and defense, variation simply remains variation.

The effect of genetic assimilation is to make expression of an innovation more reliable and more precise. If the innovation—a new organ, defense, metabolic pathway, or behavioral routine—is beneficial under commonly encountered conditions, its expression should come under the control of the genetic machinery. If the specification is too rigid, the ability to adapt to new conditions will be compromised. The combination of adaptability and precise control is most likely to be found in situations where food is plentiful and where enemies exert strong selection. The answer to the second question is thus that controlled adaptability is nurtured in the competitive marketplace of life, where opportunity for innovation is possible thanks to a productive natural economy.

I use the words *innovation* and *marketplace* deliberately for two reasons. First, they draw attention to the reality that organisms live in an economic world of trade, opportunities, and challenges, a world ruled by competition. Second, they emphasize the parallel between evolutionary improvement and the process of bringing a product from prototype to commercial success. In its early stages of development, the new product is not yet exposed to competition from other similar products. Only when an inventor has worked out the most obvious imperfections can the new item be ushered into the unforgiving competitive environment in which commodities are bought and sold.

The third question—how a rule-bound system can achieve flexibility—requires an answer that combines safety with danger. In their book *Animal Architects,* James and Carol Gould suggest that safety is a necessary condition for the evolution of behavioral innovations that entail flexibility and intelligence. New states, or the potential to create them, must arise in the developing organism. It is therefore the safety of this most vulnerable life stage that is particularly important to the evolution of new, imperfect behaviors and aspects of form. Such safety is found in well-guarded nests, termite mounds, beehives, ant colonies, beaver lodges, and burrows, as well as in the many plants and animals in which care and provisioning of the offspring occur internally in the parent's body, as in the seeds of flowering plants and in live-bearing mammals.

In contrast to the freedoms associated with life in an underexploited environment, the permissive environment of an embryo developing in the safety of a house, colony, seed, well-provisioned egg, or womb is itself an evolved adaptive response to a dangerous world. Embryonic protection and safety in houses thus neatly combine the freedom necessary for the nurturing of traits that require substantial improvement before becoming useful with the sophisticated regulation that is essential for constructing complex adaptations in a biologically rigorous environment of enemies.

The more nuanced view of the basis of adaptive innovation that emerges from these investigations is that many potentially useful novelties begin as plastic, nongenetic, unregulated responses to the environment. Even if a genetic mutation causes the initial change, it is strongly influenced by nongenetic circumstances. In the course of evolution, the innovation becomes enshrined as a gene-based trait, whose expression depends not only on the founding mutation but also on the regulatory actions of other genes. The expression of genes and of the traits they encode remains sensitive to the environment, including the environment of the developing embryo and conditions within the adult body. Genes and environment interact to produce the physiology, form, and behavior of living things.

These sequences in evolutionary adaptation find striking parallels in the history of human legal codes. Unwritten social customs or verbal agreements that are adequate in small societies, in which most individuals know one another, are typically replaced in larger human groups by formal, written laws, contracts, and—more bureaucratically—by a raft of detailed regulations, records, legal precedents, and systems of financial accountability. The invention of writing is equivalent to the imposition of genetic regulation. To some extent, at least, the resulting routines become increasingly independent of the particular situation in which they are applied, and thus become a more predictable component of the environment in which they operate.

Debates have long raged about whether genes or the environment—nature or nurture—is more important. This is the wrong question to be asking. Clearly, both genes and direct environmental effects matter in the expression of such traits as intelligence, susceptibility to disease, athletic ability, growth form in plants (tree, shrub, vine, or herb), and body size, among thousands of others. Genes and environment are not mutually exclusive or separable alternatives; they work together to create life and its adaptations. The question that is more worth asking, and that remains little investigated, is this: under which circumstances does genetic determination become so rigid that environmental influences on variation wane? I believe part of the answer lies in how much evolutionary "wiggle room" is available in a population. In a saturated world of limited resources, there is little leeway for adaptive maneuvering. Where resources are underexploited, however, relaxed genetic or legal regulation suffices, and direct influences of the environment are more apparent.

Disconcerting implications for the two principal economic systems that modern human societies have invented flow from these ideas. Few subjects have spawned more rhetoric and more mythology than capitalism and its opposite, a command-and-control economy often labeled as socialism or communism. Most Western economists are convinced that capitalism is intrinsically better than a more centralized and more regulated economic system not only because it is more consistent with human nature and has led to a higher standard of living for most people, but also because it is thought to mirror the competitive free-for-all prevailing in nature. Writing of capitalism, the American entrepreneur Michael Rothschild claims that "If the workings of the economy parallel the functions of the ecosystem, if organisms follow the same principles of form and function that govern the evolution of organisms, then there is but one natural mode of economic organization."[1] This conviction stems from the notion that regulation of economic activity, which in human society is mainly imposed by the centralized state, is absent in the economies of nature. A free-market economy, in which individuals are free to compete and trade in an unregulated (or deregulated) environment, is thus believed to resemble the economic system that has been sustained over billions of years on the prehuman Earth. Ecosystems function well despite this alleged absence of regulation, much as capitalist economies are supposed to benefit their members through the mysterious action of what Adam Smith, the Scottish founder of the modern field of economics, called the "invisible hand."

By "invisible hand," Smith meant the diffuse, unintentional, and benign regulation of the marketplace brought about through fair competition among diverse self-interested parties. Fair competition thrives as long as interest groups, including the state, have not distorted the prices of goods by imposing tariffs and other restraints on trade—that is, as long as monopolies are prevented from gaining a stranglehold over the economy. As the examples in this chapter show, there is plenty of this kind of diffuse regulation in nature: it evolved in the genetic system, in the physiological regulation of the body's internal state, and in all known ecosystems. But the natural processes of competition and cooperation have also spawned central command-and-control systems, as when centralized sensory and motor functions become concentrated in the evolving brain. Deliberate integration and coordination of functions through cooperative arrangements make Smith's "hand" visible in nature just as they do in modern economies dominated by corporate and government bureaucracies.

Nature, in other words, has produced systems that conform to the capitalist model as well as arrangements that more closely resemble a centralized command-and-control society. The question we should be asking is not which of these systems is superior, but under which conditions the extreme versions—the free-market economy and the central controlled model—work most effectively. Unregulated free-market capitalism, based on uncontrolled exploitation of resources that are for all intents and purposes infinite, is like the rapid-growth, high-fecundity strategy of early flowering plants, and early mammals: it works well as long as resources are there for the taking and as long as the economy can grow and expand in the midst of continuing plenty. Central regulation arises when competition intensifies and cooperation is required for further exploiting resources or for creating new ones. The visible hand of control emerges when a system reaches a plateau, and when the ability to innovate the economy out of stagnation requires costly sacrifice. This plateau need not be absolute; a change in circumstances may tilt the system back into a state of growth and renewal. For our species, however, politicians and the public may have to accept the possibility that, as we reach the point where the world's productive capacity can no longer grow in the size and per-capita wealth of the population, there will be a shift toward a more centrally organized economy and government. Under such a system, conservatism—the maintenance of the status quo—will more often be chosen as the solution to problems than new costly reforms.

To avoid such rigidity, a highly regulated system must incorporate ways of responding rapidly and flexibly to unanticipated situations. For organisms, the genetic system must be complemented by an infrastructure that can react quickly. A large, complex genome of tens of thousands of genes is too small and far too slow to anticipate or respond to the pathogens and environmental signals that the organism encounters in everyday life. In animals, the immune and nervous systems, composed of thousands of cells of just a few kinds, have evolved to respond quickly and flexibly to the conditions in which individuals find themselves. Although the basic components of these systems are encoded in the genome, these components interact and combine in myriad, autonomous ways, creating a nearly limitless array of states that can deal with all sorts of challenges, from new germs to threats from predators. It is the combinatorial potential of these subsidiary systems rather than direct specification of all possible responses by genes that makes rapid and varied reactions to changing conditions feasible.

Legislators and government officials might draw some useful lessons from these evolutionary insights. If, as I intimated earlier, an organism's genome is analogous to a society's code of laws, and each gene-based adaptation represents an individual law or regulation, then the most effective written statute would apply to many situations and retain flexibility. Rigidly deterministic traits tailored to specific challenges may work well for a species in the short run, but they become ineffective or even harmful as the target challenge is eclipsed by a new danger. Evolutionary adaptations are most enduring and most likely to be conserved with modification when they confer flexible and rapid yet regulated responses. This principle should apply to the legal code as well. Laws, it seems to me, should be relatively few in number, be written broadly to encompass many situations, and be enforced (or, in evolutionary terms, expressed) flexibly.

The pathway from *Littorina* to the law may seem long and tortuous, but it is emblematic of the process of adaptive evolution. Adaptation consists of a search among alternatives for an adequate solution to, or hypothesis about, an environmental challenge or opportunity, followed by a flexible codification of that hypothesis so that the solution can be preserved and transmitted. There is no reason to believe that this sequence of events should occur with the fewest possible steps, or in a straight-line march from the ancestral to the descendant condition. There are twists and turns, even brief reversals and setbacks, and long periods when nothing happens. But as long as life and environment work together, there will be adaptation, in which heritable instruction complements quick, flexible responses and sensitivity to the environment.

The key to adaptation, then, is things working together, creating structures and properties that were not there before. Life itself is a state of complexity brought about by simpler components coming together to make something truly novel. Natural selection is a process in which agencies of destruction and opportunity sort among alternatives according to how well those alternatives work, but without considering where the complexity that underlies these alternatives comes from, we cannot fully understand evolution. It is to complexity—value added when parts come together as wholes—that I turn in the next chapter.

CHAPTER SIX

The Complexity of Life and the Origin of Meaning

On a cold Thursday evening, Edith and I sit in the cavernous Hooglandse Kerk in Leiden, the Netherlands, immersed in sonic splendor, and moved to deep contemplation as the sacred music of Tomás Luis de Victoria and other Spanish late Renaissance masters, performed alternately on the great Baroque organ and a capella by the choir, fills the church. Everyone and everything—the audience, the chords and melodies, the organist and his craft, the singers and their conductor, the artistry of the composers, and even the church with its echoing acoustics—have become one, a three-dimensional edifice of harmony and meaning. It is a transcendent construction, inspired by a fervent faith in God, anchored in beliefs in miracles and creation stories and the afterlife, all doctrines I rejected long ago. But the magic—the ecstasy created by coordinated complex machines inside a massive box of stone—endures. The components taken singly may be ordinary and undistinguished, but the whole achieves a grandeur and significance, a richness of experience, that transcends the context in which it was created and that makes life more than mere metabolism. And all this splendor—the material parts as well as the spiritual dimension—is made possible by a process of evolution that began with lifeless molecules, devoid of all meaning, engaged in chemical reactions in environments that we would regard as profoundly alien and hostile.

This is as compelling a demonstration as can be found of parts working together to produce an emergent whole, a unit with properties that none of its components possesses. Water, a molecule consisting of two hydrogen atoms and one oxygen atom, is utterly unlike the two component elements. It is a liquid rather than a gas at room temperature, it expands rather than contracts in the solid state, and it is an exceptionally good conductor of heat. The properties of water seem irreducible, much as our complex brain might appear to be irreducible to its many constituent parts; but in fact they arise through the interaction—the working together, or synergy—of components. Likewise in music, chords and melodies convey patterns and evoke emotions that single tones cannot. Sentences, paragraphs, and books have meanings that individual words and letters do not. Living things, too, work together to add dimensions of value, function, and meaning. Survival and propagation are themselves expressions of emergence and synergy common to all life-forms; but we humans are motivated and enriched by more than these lifewide aspirations. We perceive a greater purpose—through love, curiosity, a social conscience, helping others, and perhaps above all, through aesthetics—a deeper meaning that makes our individual lives worthwhile to others. Without that added significance, and without the intentionality that enables us to create a future according to our tastes and values, life would be empty; we would descend into apathy and callousness. Purpose and meaning, however they come into our lives, are as real and as essential as the evolved imperative to survive and reproduce.

Where do these particular human values and aspirations come from? What is the origin of the intentionality—goal-directed action that fashions a desired future—which we share with other forms of life? How can these expressions of synergy—the phenomenon that lies at the foundation of all complexity—be reconciled and explained in an adaptive evolutionary framework whose building blocks are ultimately inanimate and whose construction proceeds through unintentional combination and selection?

Few concepts in science are as difficult for people to embrace as the idea that complex life, and especially humans and their remarkable life of the mind, can arise through the interactions of everyday agents without direction from some guiding intelligent force. Even if people could accept the unintentional emergence of water from interactions between two kinds of atoms, or the natural formation of rock from mineral components, life is so special and so apparently irreducible that emergence through natural means seems preposterous. Moreover, skeptics often express the profoundly disquieting fear that the cold materialism of chemistry and random chance deprives life of all higher purpose and meaning. For them, only an all-powerful intelligent creator and overseer can imbue mere existence with moral direction and with a sense of right and wrong. Without such an agent operating above the human fray, so the argument goes, life would be pointless and chaotic, little more than an endless parade of chemical reactions in an uncaring, amoral, if not evil universe where aimless selfishness and greed reign. So if the higher faculties of human mental life demand the intervention of someone or something supernatural, then evolution, insofar as it is an unguided process of variation and selection, cannot offer a sufficient explanation of life's complexity or for our values. Indeed, given its implied rejection of supernatural agency, evolution is often tarred as alien and evil, as corrosive to the established moral order, as a dangerous ideology that tolerates and even invites unacceptable behavior. Evolution in this view threatens society by reducing human existence to a molecular maelstrom of metabolism and moral indifference.

At the level of rational thought, these attitudes stem from the intuition that the discontinuities we perceive around us—between life and nonlife, between intentionality and instinct, between complexity and simplicity—reflect real gaps that natural processes cannot bridge. The states we associate with life and with being human appear irreducible, requiring the existence of an intelligent, unknowable agency that eludes scientific investigation. Where such a force would come from is left unanswered—indeed unasked—as is the question of how that agency—a deity, prime mover, or what have you—goes about creating the irreducible. A related and equally entrenched belief is that the human species and its institutions stand apart from all other manifestations of life. Evolutionary explanations might suffice for small changes of the kind observed when wild plants and animals are domesticated through selective breeding, but they are judged to be inadequate and inappropriate to explain our uniquenesses.

Given these widely held viewpoints, it is essential to come to grips with how known, observable, natural phenomena can explain how the great transitions that gave us life, social behavior, and human faculties of the mind came about. With this scientific understanding as a foundation, it then becomes important to show that the values and purposes we rightly cherish are no less wondrous and essential when they emerge from natural evolutionary processes as when they are magically handed down from above by an inscrutable "intelligent designer." As the American evolutionary biologist and biochemist Stuart Kauffman observes, "Evolution is not the enemy of ethics but its first source."[1]

Transitions from one state to another are so common that we rarely give them much thought. Ice, liquid water, and steam have radically different properties, yet they represent different states of the same simple molecule, H_20. As the temperature rises, it crosses particular thresholds—the melting point of ice and the boiling point of water—at which there is a transition first from a solid to a liquid, and then from a liquid to a gas. As is also the case during the processes of condensation and evaporation, the changing energy states of the molecules bring about dramatic shifts in the ways in which molecules interact with one another. The relatively weak bonds between the hydrogen atoms of adjacent water molecules keep H_20 in the liquid phase but loosen their hold when molecules speed up in the gas phase. Tipping points, or phase transitions, occur when one kind of interaction comes to predominate over another in a system of particles that by virtue of their random motions exchange energy. Interaction is the foundation of all emergence and all synergy; it is the agency that causes simpler components to combine and work together, forming new structures with new properties that seem irreducible but that are, in fact, the consequence of how particles move and exchange energy.

The centrality of interaction as the agency of complexity is well illustrated by the origin of life from nonlife. In the view of Stuart Kauffman and many other scientists who probe life's origins, life represents a phase transition of complexity, the product of an unintentional, collaborative enterprise in which lifeless molecules interact so frequently and with so many other molecules that they form a network. This network becomes the autonomous, self-sustaining, replicating state we call life. On the early Earth, according to current theories, simple common carbon-based molecules such as nucleotides and amino acids each represented by several variations on a common molecular theme combined to form chains, or polymers, that react with other polymers. Outside energy is required to drive many of these reactions, but the flexibility of the polymers—ribonucleic acids (RNAs) composed of nucleotide bases and proteins composed of amino acids—lowers the energy hurdle that must be overcome to complete the reaction by stabilizing the often bent and stretched chemical bonds that form during the course of the reactions. The polymers, especially chains of amino acids but also RNA itself, act to speed up, or catalyze, the formation of molecules other than themselves.

Because of the reduced energy costs, catalyzed reactions proceed faster than reactions in which the energy barrier remains high. The chance that a reaction is catalyzed increases as the pool of polymers grows larger. Once a critical number of molecular types is reached, and there is at least one catalyzed reaction for each type of molecule, a phase transition takes place, forming a catalyzed network of interacting molecules. Some of these networks form self-sustaining loops, or autocatalytic cycles, in which chemical reactions that require energy are coupled with reactions that release energy. Such self-sustaining networks have entered the living state. The chemical details of how such a network comes about remains the subject of a vigorous scientific debate and investigation, but the basic principles—phase transition, catalysis, and self-reinforcing couplings between energy-demanding and energy-producing reactions—are not in doubt.

This series of events may not have occurred always and everywhere on the early Earth, but under suitable conditions, the transition to life and the formation of life's cellular organization may have taken place on many occasions. A unique origin of life therefore seems unlikely; it is inconsistent with the expected emergence of autocatalytic networks in energy-rich environments like volcanoes, where a diverse array of simple organic compounds on mineral surfaces and an abundance of metals like iron provided ideal conditions. Likewise, the double-layered membranes of fatty acids that surround and attach to the molecular machinery of a living cell can form spontaneously when the fatty-acid molecules orient themselves with their water-seeking ends facing the cell interior and the outside environment while the water-avoiding ends point toward each other within the membrane. In Kauffman's words, "Life is an expected, emergent property of complex chemical reaction networks."[1]

A process similar to natural selection among living individuals emerged early during the molecular events leading to the state of life. Not only were some polymers more stable than others, and therefore more likely to persist, but some reactions, especially the catalyzed ones, were more likely to take place than others. Differences in persistence and reactivity likely existed long before the specialized machinery of replication of RNA and DNA had evolved, and long before individualized life emerged. In short, life is an emergent state, made possible by the working together of diverse molecules in an energy-rich setting. The particular manifestations of life, ranging from cells to ecosystems, succeed because differences among molecules, among cells, and at other levels of the organization have become associated with function—that is, with how well they work.

One reason why it has proved so difficult to reconstruct the origin and history of life on Earth is that living things as we know them today, some three-and-a-half billion years after the first appearance of life, bear little resemblance to ancestral forms. The organisms we see today are arrayed at the tips of branches on an evolutionary tree. Their properties, including traits like the genetic code that we take as universal, have been honed by selection for eons, during which the conditions of life have also changed dramatically, thanks in part to the work of life itself. We must therefore shed our intuition that the present is an infallible key to the past. Few of the ancestral states of life are likely to have persisted to the present day, and those that have persisted may linger in conditions quite different from those in which they originated. For example, organisms for which oxygen is a lethal poison are today found only in places like sulfur-rich sediments of salt marshes, decaying meat, and other environments where free oxygen is absent; but in the days before oxygen became widespread on Earth's surface (beginning about 2.4 billion years ago), such organisms would have existed wherever life was feasible—on rocks, in sand and mud, in surface waters of the ocean, as well as in sulfurous habitats.

Historians of human affairs know the bias of current conditions on the interpretation of past events by the awkward label "presentism." The good historian knows that the beliefs and actions of people in the past must be interpreted in the context of historical circumstances, not according to modern sensibilities. The same principle applies to the work of evolutionary biologists. We must be prepared to accept the reality that the form, behavior, physiology, and complex molecular machinery of today's organisms always represent highly derived states, which are separated from the ancestral states by vast stretches of time and billions of evolutionary steps.

Take, for example, the evolution of the genetic code. In nearly all living organisms, nucleotide triplets in RNA and DNA consistently correspond to particular amino acids in proteins. But this universal genetic code represents a derived state. Molecular biologists grappling with the origin of this code have adopted the view that genes (sequences of DNA and RNA) coding for small proteins composed of just a few amino acids were the norm in the earliest life-forms, those that in today's world would have the organization of bacteria and Archaea, the two most basal branches on the evolutionary tree of life. Instead of being transmitted vertically—from one generation to the next—these gene sequences moved frequently and easily between cell lineages, rather like words or sentences exchanged between people. The pool of available gene sequences was not integrated into individual genomes, as in living organisms, but instead was the property of a community of cell lineages. By sharing and combining genes from this pool, early life-forms created longer chains of genetic material, which in turn specified longer, more complex proteins, including enzymes. These enzymes in turn catalyzed a host of chemical reactions that were necessary for making and breaking down energy-storing substances like ATP (adenosine triphosphate), which were themselves crucial for the work of life. By taking advantage of the power of combination, the community of life constructed the biological infrastructure of metabolic and synthetic pathways that all living things now see.

In this genetic web of lineages, the genetic code is rather like a common language—a lingua franca, if you will—enabling lineages that differ in composition to exchange genes and to build new pathways and structures. Any departure from the code that might have arisen in one lineage in this collaborative community would have been quickly eliminated by selection, because genes specified by such deviant codes would be unable to propagate or to be recognized. Deviations from the universal code are known, but they occur only in a few evolutionarily more derived forms of life, such as yeast, where gene exchange between distantly related lineages is infrequent. The rarity of deviations from the universal code minimizes the harmful effects of mutation and mistranslation. The combination of extensive sharing of genetic information and intense selection against deviations from the common language is responsible for the establishment of the code as we know it today.

Exchange among lineages, or horizontal transmission, may be less common in plants and animals than it is and was among lineages of simpler life-forms, but it is still extremely important, especially because it creates variation and therefore provides the raw material for innovation and selection. For example, horizontal exchange of genes is the hallmark of sexual reproduction, which characterizes the vast majority of plants, animals, and fungi. A daughter genome is formed when genes from the two parents are recombined and then segregated in such a way that it differs from the genome of either parent. The Austrian marine biologist Wolfgang Sterrer has intriguingly suggested that scrambling of the genome during sexual reproduction may be an evolved defense against cellular parasites such as viruses. If intact genomes were transmitted vertically from one generation to the next, then a virus whose genome has been inserted into that of its host cell would itself be faithfully inherited; whereas with recombination, either the virus's genome is dismembered or the virus finds itself in a new, potentially hostile genetic environment in the offspring. In sexually reproducing organisms, horizontal gene exchange is for the most part between closely related individuals, but exchange among organisms that are widely separated on the evolutionary tree is also known. Some genes of nitrogen-fixing bacteria have become incorporated into the genomes of their plant hosts, with the result that the partnership between the parties is a highly integrated association of mutual dependence. It is in fact likely that the genomes of all plants and animals contain elements derived from "domesticated" microbes.

From its beginnings, then, life represents a form of unintentional collaboration, the mutual, catalyzed interaction of diverse simpler elements, which are themselves not alive, to yield a novel, dynamic state capable of doing work as expressed in biochemical synthesis, metabolism, growth, replication, and movement. To sustain itself, this cooperative venture needs material and energy resources, which cycle into and out of the living body. As long as there is more than one living thing in a chemical neighborhood, there will be local competition for these external commodities. Living bodies capable of acquiring or retaining more resources hold a competitive advantage under the conditions at hand. Competition thus ultimately provides the basis, or primary agency, of selection. This property of life establishes an emergent criterion of performance of function that, together with the inevitable fluctuations of the environment, helps determine the fate of bodies and of propagating lineages.

Just as catalysts favor some reactions over others by establishing a form of molecular cooperation, so living things may also benefit one another through a process of mutual reinforcement. Such associations begin as coincidental relationships that are neither permanent nor consistent, but if the benefits of the new relationship are great enough, selection among the parties involved will cement the relationship. All the world's ecosystems contain numerous arrangements of this kind, in which one species—say, a herbivore fertilizing the soil—benefits others on which it depends—plants, in this case—and in which success in the struggle for life depends not only on one's own characteristics, but also on the actions and capacities of others. If the advantages outweigh the inevitable costs of competition among the partners (for example, the costs of the plants being eaten), selection may favor the emergence of a more intimate partnership that can function as a unit of competition, because the association becomes both permanent and obligate.

Although these relationships often involve members of lineages that have long been separated and that occupy different positions on the evolutionary tree, as in the example of herbivores and plants, others are social groups of individuals belonging to the same population. Being social means that, despite conflicts and diverging self-interests, individual organisms cooperate in, and belong to, relatively permanent groups, which can achieve ends that individuals cannot achieve if they act alone. Individuals benefit because, by pooling their capacities and communicating among themselves, they create a new whole with greater power and a wider reach; they achieve safety in numbers and a collective advantage in the never-ending struggle to acquire and retain resources. An integrated group has more sensors detecting signals from more places and over longer durations than any individual member. If even one member detects danger or food, effective communication enables the signal to be transmitted quickly throughout the group and appropriate responses to be mounted. With division of labor, in which classes of individuals take on specific functions—defense, synthesis, feeding, tending the young, reproduction, and so on—a highly integrated group can defend itself and its members against more potent foes, and can more effectively locate and exploit the necessities of life. Group hunting as observed in wolves, African hunting dogs, and even some social spiders raises the success rate of prey capture and allows larger prey to be secured. In other words, being social is highly effective in settings where enemies pose potent danger.

Social or colonial species in fact dominate almost all environments on land and in the sea. In the tropical rain forests of South America and Southeast Asia, social ants comprise more than 80 percent of the biomass of all arthropods. Social predators—humans, whales, wolves, and various fish—are the top consumers in their respective ecosystems. Flowering plants—the dominant land plants of the last one hundred million years or so—depend on the efforts of pollinating social bees and bats for their extraordinary success. On the seafloor, communities of sedentary animals are everywhere dominated competitively by colonial types, in which "individuals" are physically connected, all being derived through asexual reproduction from a single founder. Familiar examples are reef-building corals, moss animals (bryozoans), and colonial sea squirts.

Societies of the kind seen among insects—ants, bees, wasps, termites, and some aphids—are composed of individuals whose genetic compositions are either identical or extremely similar. All individuals in a colony are derived from matings of a single pair or a few pairs, much as the cells of an individual are all derived from a single fertilized egg. The society thus functions as a kind of superorganism, an entity in which the parts are dispersed instead of physically connected into a single body. The dispersed members thus have closely similar genetic interests, borne out of their close kinship, so that there is little conflict within the group. Group identity and cohesion are readily achieved. Given this close kinship, individual sacrifice, including suicidal roles in defending the group or its founding pair, can arise because each member contributes to the success of its own genes by helping its kin when acting on behalf of the group. Extensive research on ants has shown that intense competition among groups—often aptly referred to as warfare—enhances altruistic behaviors of individuals within groups. It may, in fact, be this intense competition for territory and food that pushed insect societies through the phase transition between a relatively undifferentiated commune to the caste organization of functionally specialized members that we see today.

Humans have achieved group identity differently. Largely through cultural means—a common language, shared religion, nationalism, distinctive dress codes, and a host of other social conventions—genetically distinct individuals have banded together into groups that vie for resources, often expressed as territory. Disputes among groups, ranging from raids to total warfare, amply testify to the competitiveness of groups as well as to the cooperation—often among strangers—within groups. It isn't just groups organized to fight over territory and its resources that give human civilization its unique character. People organize themselves into groups of all sorts—governments, orchestras, school systems, universities, monasteries, corporations, hospitals, charities, labor unions, professional societies, churches, and sports teams—which accomplish tasks and perform services to society at large that no single individual is able to do.

In our species, the variety of groups has far surpassed that in other animals, and the traits that enable individuals to form stable groups functioning as emergent units uniquely define us as human. Groups reward cooperators and punish selfish behaviors by means of elaborate expectations, codes of conduct, formalized laws, and customs. The effect is to increase the benefits of membership and the costs of pursuing conflicting self-interest. The rewards are many and varied; sometimes they are financial, but perhaps more often they are expressed as reputation and status in the group. Many of the values we esteem as virtues—valor, honor, fairness, benevolence, trustworthiness, and empathy, among others—contribute directly to the cohesiveness and effectiveness of groups. So do patriotism, religiosity, and the tendency to conform. But if groups are to function as potent competitive units, they must also coordinate and reward behaviors that would be highly destructive if used within the group. Hatred of outsiders, violence, ruthlessness, envy, jealousy, and callousness are just as human as are the traits we uphold as virtues. In a social species like ours, in which stable, cohesive groups compete as units and have their own emergent cultural adaptations, individual traits that benefit cohesion within the group and competitiveness among groups are strongly favored.

One question that has preoccupied evolutionary biologists as well as students of human society is how individuals with different self-interests can be made to cooperate and even to sacrifice some of their advantages in order to help others and to benefit the good of the group. An appealing answer, though in my view an insufficient one, lies in the phenomenon of kin selection, in which an individual is more likely to sacrifice for a very close relative than for someone who is unrelated. The idea is that the sacrificer is helping to propagate his or her genes when that individual is helping kin but not when he or she is being a good Samaritan to a stranger. According to this view, the most cohesive, and therefore the most effective groups are those composed of close relatives, because an altruistic act in favor of the group is not really inconsistent with individual self-interest. What is good for the group is also good for the individual members.

The problem with this theory is that many highly integrated and well-functioning groups achieve cooperative harmony even though members are unrelated to one another. In the human realm, it is the cultural glue that holds such groups together and that encourages self-sacrifice even among strangers. Military fighting units work because individual soldiers develop bonds of friendship and allegiance with fellow soldiers and because the units stay together long enough for such ties to be strengthened. But nature, too, is replete with groups and partnerships in which the members are unrelated. Though not intentionally created as are human groups, symbioses involving multiple parties have evolved countless times and have persisted for millions of years without the benefit of kin selection among their constituents. The advantages of forming an emergent alliance among divergent creatures must be so large that selection in favor of the partnership as a powerful unit of competition and defense overwhelms selection for conflicting self-interests of the members. Close kinship surely makes cooperativity and self-sacrifice easier, but it is not essential. "The enemy of my enemy is my friend," as the saying goes; and this pragmatic reality in the face of a common threat is a powerful agency for bringing together competing individuals to form an effective emergent entity whose competitive powers exceed the sum of their members' levels of performance.

Long-distance trade among strangers has been claimed to be a uniquely human form of cooperation, along with the institutions that make it possible. Trade among strangers, especially when the parties are far apart, requires mutual trust, and this trust, which reduces risk and enables investment, is engendered by banks, stock markets, legal contracts, patents, and a judicial system of laws and enforcement. In short-distance trade and in small societies, frequent interactions among the same parties may be sufficient to nurture trust, but more formal mechanisms, including government, are needed to curb cheating and to protect the economic interests of trading partners when the partners are mutual strangers. While these institutions do seem to be uniquely human cultural inventions, all except government appearing in the last thousand years of human history, trade and other forms of cooperation among unrelated individuals are deeply rooted in adaptive evolution. Humans have simply taken the principles of symbiosis from the scale of intimate contact and genetically prescribed rules of engagement to a geographic scale and culturally formalized understandings.

Although much of the strength and singularity of the human species is due to our social nature, individuals have, throughout history, played decisive roles as originators and innovators. Individual initiative has been crucial to the founding and leadership of states; the origins of the world's major religions; all the inventions and major scientific discoveries; the establishment of great economic enterprises; and the masterpieces of art, literature, and music. Society's role has been twofold: it provides the necessary infrastructure that gives individuals the freedom and skills to work, and it is instrumental in deciding whether a leader or innovation becomes accepted or adopted. In the most competitive societies, creativity finds fertile ground and shapes the cultural landscape. Humanity is unique not only because of self-sacrifice in favor of cooperation, but especially because of the strong positive reinforcement between society and the individual. We combine the benefits of group action with the engine of individual genius.

Are there unbridgeable gaps between life and nonlife, or between human existence—meaningful, individual, purposeful, and inventive—and what looks to us like the desolate slog of nonhuman life, with only survival and propagation as a foundation of meaning? I think not. In the beginning, life was, in Daniel Dennett's evocative words, an "impersonal, unreflective, robotic, meaningless little scrap of molecular machinery."[2] With the evolution of sensory and motor systems and of cooperation, many animal lineages became endowed with emergent emotion, purpose, and preferences. Shared culture provides the means by which the diverse forms of utility—happiness, pleasure, dignity, and respect—were molded in our social species.

Our values and preferences influence how we conduct our lives, what we buy and sell, and how we engage with others. Even if these choices are often based more on emotion than on rationality, they are conscious and intentional, with an expected outcome firmly in our minds. The calculus of our decisions and actions expresses how, with a combination of reason and our state of mind, we assess and influence the environment, including our social relationships and our future.

More than any other human attribute, intentionality—the ability to anticipate, modify, and create outcomes in designed ways through behavior involving forethought—has been held up as imparting to us entirely new capacities of manipulation and prediction. The gulf between conscious, strategizing humans, who choose and decide with specific goals in mind, and automatonlike "instinctual" creatures is perceived as wide and deep. Although it is notoriously difficult to peer into the minds of other animals, we now know that simpler versions of many of our mental traits and social norms exist in other species—some close to our hominid ancestors, others very far removed from us on the evolutionary tree. There are substantial predispositions to intentionality and a moral sense in primates, carnivores, elephants, cetaceans, and beavers, as well as in such large-brained "smart" birds as corvids (crows and jays) and parrots.

Apparently goal-directed behavior is displayed by beavers when they construct dams. Beavers live in family groups in large chambers, some up to a diameter of two meters. The chambers are above the level of the water, but they must be reached by swimming under water. Where the water level naturally fluctuates, as in many streams, beavers must ensure that their lodges remain above the water. To accomplish this, they cut and float timbers, haul in large boulders, and arrange these components to form a dam, which diverts flow and creates a pond with a constant water level. In the course of constructing the dam, beavers may dig long canals into the forest in order to gain access to more timber. They also repair damage and engage in other maintenance activities. None of this work is stereotyped; it is instead creative and flexible, all with what appears to be an abstract representation of an effective ideal dam.

There are many other examples of intentional behavior. Several species of crow have developed standardized methods to manufacture one-piece tools for extracting food from holes and other places that would otherwise be impossible to reach. Jays cache nuts with the intention of consuming them at some future time. Aware of the possibility that the food could be stolen by other jays, observing where the nuts are being hidden, the Western scrub jays in our garden seek out sites where they will not be observed by onlookers. Some will even mislead their rivals by burying stones instead of nuts. These birds show levels of awareness, of causal reasoning, and of flexible responses comparable to the great apes. They can generalize from particular circumstances, plan ahead, and engage in abstraction and forethought. If these birds could carry objects with their feet, as parrots do, they might well have been suitable ancestors for social descendants with humanlike cognitive abilities.

Dolphins, elephants, and chimpanzees—representing three distinct evolutionary branches of mammalian evolution—are among the animals that show evidence of assisting one another, as people do. Food-sharing, a human behavior that promotes greater equitability among members of a social group, is relatively widespread in mammals, including some carnivores (wolves, hyenas, and mongooses), vampire bats, and nonhuman primates such as gibbons, marmosets, and capuchin monkeys. It is therefore a trait that has arisen repeatedly in many lineages in many places. A sense of right and wrong, or at least of fairness, exists in dogs. Although no animals other than humans possess all these components, predispositions to our cognitive and moral world are widespread on several branches of the vertebrate evolutionary tree.

Even the nonsocial octopuses show traits of self-awareness, consciousness, and individual preferences that we associate with the cognitive attributes of large-brained apes and birds. Recent work by Canadian biologist Jennifer Mather and her group has revealed substantial cognitive flexibility in species of *Octopus* with respect to penetrating the shell defenses of prey clams. As shown by the observation that octopuses do not retrace the outbound path when they return from a bout of foraging, these cephalopods have an internalized map of their surroundings as well as a flexibility of response that is otherwise unknown among molluscs. There is even tool use: shelters are constructed with the aid of the water jet—used by other cephalopods in backward locomotion—from the flexible funnel of the animal's mantle. Highly developed abilities to change skin color and pattern extremely rapidly on very small areas of the body hint at sophisticated communication between individuals. Traits converging on the cognitive abilities of social species, such as our ancestors, can therefore arise even in solitary animals that are very far removed from us in the animal kingdom.

Mounting evidence from behavioral experiments on and observations of other animals indicates that the distinction between intentionality and unconscious, short-term selection is fuzzy at best. Although hunger and the drive to reproduce might seem to be barbarous values in comparison to our supposedly more refined ones, they are nonetheless values, much as survival is. Knowingly or not, organisms do things that are in keeping with these values. Anger, shame, pleasure and other emotions are well known in vertebrates; and whether the aggression of a wasp is conscious or instinctual given the right stimulus is of little concern to me if I am about to be stung. In his original conception of natural selection, Darwin looked to the breeding of domesticated animals as an informative model for selection and adaptation in the wild even though domestication proceeded with intentional human choices, whereas natural selection may be imposed by agents with no expectations or hopes for the future. Directed selection such as domestication may yield characteristics that are not normally found in the wild, such as drooping ears in dogs, but natural agents of selection operating without foresight likewise elicit one-of-a-kind adaptations that we find bizarre. Sexual selection has produced outrageous visual and acoustic displays in birds of paradise, cannibalism of male spiders by females, male copulatory organs that penetrate the skin of land-snail partners, elaborate hornlike excrescences on the heads and bodies of beetles, and flowers with lovely scents.

In his excellent book *The Tinkerer's Accomplice,* the American physiologist J. Scott Turner argues persuasively for a very broad evolutionary distribution of internal body states that, when looked at in hindsight, give the appearance of having been intentionally created. An example of this is warm-blooded birds and mammals, which maintain a high, nearly constant body temperature despite wide variation in the temperature of their surroundings. Everyday activities such as digestion, respiration, blood circulation, brain function, growth, and cell division proceed in a highly regulated internal environment that, thanks to innumerable physiological feedbacks, differs dramatically from conditions outside. The stable internal environment creates predictability, and therefore unintentionally maps out the most likely future states, in the same way that we map our futures through directed deliberate actions.

Skeptics might claim that, even if seamless transitions between animal intelligence and human cognition can be demonstrated, the symbolism, conscious intentionality, and moral sense that define our mental makeup introduced capacities that set us apart from all other animals, including our immediate hominid ancestors. Humans alone play in orchestras and are transported to heights of introspection while listening to a great choral work. We alone learn enough about the universe to build atomic bombs, forecast the path of the next hurricane, and piece together our own history. And we alone can acquire a house, a piece of land, or a bottle of rum by handing over a few sheets of paper, bits of metal, or a plastic card, all means of exchange built on little more than trust. With these evolutionary departures, we humans have been able to expand life's horizons and to create our future as no other species has been able to do. But I believe that the creative combination and learning that lie at the root of our unparalleled power have a long pedigree. The ability to create a future has been intrinsic to living things for billions of years.

Consider the evolution of human tool use and language. As inanimate extensions of the body, tools are used by a number of birds, mammals, and even some insects to build nests and other shelters and to obtain food that is out of easy reach. These tools—probes, spears, hammers, and anvils—are one-piece implements modified from readily available objects like stones and sticks. Beginning about two hundred and eighty thousand years ago in Africa, our hominid ancestors of the Middle Paleolithic made the transition to a combinatorial style of toolmaking. The new compound tools consisted of three or more parts—initially a handle, a functional device, and a binding between the two—that must be manufactured separately before they are put together to form a new functional whole. The repertoire of materials and parts expanded episodically, so that by the time the modern human species arose some time after two hundred thousand years ago, the diversity of tools increased dramatically, to include nets, traps, baskets, and pots, among many others.

In parallel, the unitary utterances that our ancestors likely employed for communication have in the modern human species been extended to complex strings of tones and eventually symbolic notation. Separate elements—tones, rhythms, and different configurations of the mouth cavity made possible by a tongue deeply inserted in the back of the mouth—are brought together in open-ended combinations to form chords, melodies, words, speeches, and written documents. Our combinatorial language carries symbolic meaning, representing things and concepts and states of mind. Animals also convey information through sounds or other displays—the songs of birds and whales, and the dances of bees returning to the hive from a scouting expedition in search of nectar—but the components are stereotyped and, with the exception of thrushes and a few other birds capable of producing two notes simultaneously, in linear sequences. Humans have crossed a communication threshold into a novel cognitive realm thanks to the power of combination and the emergence of a correspondence between sequence and meaning, much as earlier transitions established life, the genetic code—another correspondence between sequence and meaning—and social organization. The details of how our cognitive transition was achieved anatomically and socially remain elusive, but the basic outlines of the transition conform to earlier transformations resulting in a vast expansion of capacities and the emergence of new states.

In short, the uniqueness of intentionality has, I believe, been exaggerated. True, conscious choice implies goal-oriented actions, but such intentionality has been approached independently in many animal lineages. Humans have unquestionably gone much further than other animals in adding dimensions of value and meaning to their lives, of anticipating and rapidly adapting to change, and of achieving at least short-term control over their surroundings, through a combination of flexible learning, abstraction, and projecting into the future. But these derived capacities confer no radically new agencies of adaptation. Intentionality surely expands the range of adaptive possibilities, especially when the goals can be realized only far into the future, but selection—intentional or otherwise—is still exercised through age-old competition and cooperation among emergent agents. Humans are at one end of a continuum of emergent meaning and control, not on one side of a steep ravine that separates us from the rest of the animate world.

The scientific idea that our values and cognitive abilities arose from simpler states in our evolutionary past directly contradicts the widely embraced religious doctrine of supernatural agency. Many of the world's major modern religions have tapped into our socially inculcated tendency to seek leadership and guidance by reinforcing codes of good behavior with an appeal to an all-knowing, all-powerful god. With threats of eternal damnation and promises of eternal bliss, as articulated by human surrogates, the supreme being's control is perceived as more potent and more sweeping than is a system of laws devised by people acting without the blessing of holy authority. The supreme god infuses mere existence with higher purpose and meaning by demanding obedience and by creating the conviction that, by living a life in accordance with rules laid down in holy scripture, people become part of something bigger and better than themselves. In the worlds of a popular Christian evangelist, "If there were no God, we would all be 'accidents,' the result of astronomical random events in the universe. . . . [L]ife would have no purpose or meaning or significance. There would be no right or wrong, and no hope beyond brief years here on Earth."[3]

As a natural historian of the human condition, I am persuaded that gods, ghosts, angels, miracles, and the afterlife exist in many people's minds. They are as real to them as scientific ideas and fossil life-forms are to me. Their power to influence people's behavior therefore cannot be doubted. But would morals and group cohesion, purpose and meaning, truly wither away if supernatural mythology were replaced by a social contract more in line with the findings of empirical science?

I do not think so. Although rules of good behavior are essential if societies are to accomplish ends that are beyond the means of any single individual, societies with a secular foundation exist in large numbers and have played pivotal roles in transforming civilization. We may disagree about the merits of their accomplishments or even about the ends themselves, just as we can justifiably object to the actions and aims of societies inspired by deities or tyrants; but corporations, trade unions, educational institutions, and many modern nation-states are secular bodies that have wielded enormous influence for more than four hundred years. Conduct that benefits groups as well as their individual members need not be sanctified as the inalterable and uncontestable decree of an oligarchy or a mystical deity in order to be effective. The key to effective secular control, it seems to me, is to channel self-interest in directions compatible with the common good, and to distribute power broadly so that the errors and narrow self-interest of any one entity do not jeopardize the larger group as a whole.

The rules that hold groups together are social adaptations that, like the adaptations we observe among living things all around us, must change as circumstances warrant. To be most effective, therefore, they must be flexible, which the laws that come from the words of a god or a dictator are often not. Religion is, of course, a social adaptation in its own right, one that has demonstrably worked well for human societies in the past. Scholars will have to assess whether secular governments or theocracies work better in the long run, but a secular loosening of controls does seem to have enabled societies to enter into periods of advances in living standards.

On a more philosophical note, a spiritually rich, meaningful life likewise seems to me not to require the mental existence of a supernatural being. Far from it. Beethoven's *Sixth Symphony (The Pastoral)* and Schiller's poem *Heidenröslein,* so beautifully set to music by that most lyrical of composers Franz Schubert, are secular expressions that enduringly and lovingly evoke the beauty of nature. They have lost none of their emotional pull or their stimulus to reflection in our analytical age. Though springing from a religious and often anti-scientific tradition that I reject, hundreds of musical masterpieces—the choral tapestries of Palestrina, the organ works of Sweelinck and Buxtehude and Bach, and Verdi's *Requiem* among them—fill me with the same sense of grandeur as do the productions of nature and the achievements of scientific understanding. Hearing such works in a great church or a modern concert hall is for me like being on a Palauan coral reef or in a California redwood forest. The wonder and contemplation that such experiences elicit are not in the smallest degree diminished—indeed, they are enhanced and enriched—by the scientific worldviews we have collectively constructed.

From initially meaningless interactions among chemical compounds, the exchange of matter and energy in catalytic loops leads first to the emergent imperative of survival and propagation—the very essence of life—and eventually the gradual appearance of awareness, purposeful action, the perception of meaning, and a desire for accomplishment, the all-important realization that there is utility in existence that transcends the ancestral, previously sufficient, goals of persistence and replication. In our species, it is the feedback between our social nature and our individuality that defines the values by which we live. For me, the evolutionary understanding of values and morals and aesthetics underscores the responsibility that we, individually and collectively, must bear for enriching our lives with purpose and meaning. It is up to us to add purpose and meaning to our lives; we cannot shift this responsibility—I would say this privilege—to an invented supernatural being whom we cannot know.

CHAPTER SEVEN

The Secrets of Grass: Interdependence and Its Discontents

When our daughter Hermine was in kindergarten, her teacher Mrs. Rudkowski thought up a creative way to make dinosaurs come alive. The children would draw the giant creatures in their natural habitats. Hermine and her classmates dutifully conjured up some version of a lumbering behemoth, but they were at a loss when it came to the beast's haunts. Mrs. Rudkowski helpfully suggested adding some grass. After all, big animals like cows and sheep graze, and therefore it seemed only reasonable that plant-eating brontosaurs and stegosaurs likewise consumed grass. Mesozoic meadows populated by megafauna—what a beguiling thought, so peaceful, yet so primordial, a premammalian paradise.

The difficulty with this imaginative intuition is that grass was nowhere to be found during the Jurassic heyday of the largest plant-eaters that have ever lived on Earth. Grass is such a familiar part of the modern landscape that a world without it is hard to conceive. Yet the geological record shows that grasses, and the prairies and savannas they create, arose very late in the history of life on land. Casting our minds back to the age of dinosaurs and beyond compels us to confront the present, and to ask questions we might never have considered. What is it about grasses and grasslike plants that makes them so successful today, and why were those characteristics absent in the world's vegetation during ancient times? Without the perspective of the past, we would never think to ask such questions.

There are larger points at stake here. The grasslike growth habit would have remained a minor evolutionary curiosity were it not for the evolution of plant-eaters with large appetites. Grasses are at the center of a web of relationships in which the participants—predators, grazers, grasses, and microbes—have come to depend on one another not just for nutrition, but also for adaptations. The grassland ecosystems that the plants, microbes, and consumers mutually created represent a geologically recent development, one in which producers and consumers have unintentionally colluded to forge a productive enterprise and an intense regime of selection. This evolution also happens to have laid the groundwork for our own grain-based agriculture. The interdependencies that make high rates of consumption possible in the world's grasslands are by no means unique to them; indeed, they characterize all living ecosystems, arising as unintentional consequences of selective regimes that are ultimately set by enemies—plant-eaters and predators, in the present case—which are perpetuated because they benefit all parties concerned. But interdependencies also confer a certain fragility to the systems in which they have evolved. The elimination of key links through catastrophe and overexploitation may initiate ripple effects that could lead to a wholesale collapse both in the pathways of energy transfer within the system and in the pattern of selection that enforces high performance among the system's members. Because mutually beneficial relationships have pervaded human society as well, an understanding of interdependency and its consequences in nature can help us grasp the opportunities and vulnerabilities that these relationships bring.

To uncover the secret of success of the grass habit, we need look no further than to that most suburban of machines, the lawn mower. When blades of grass are cut, it is their oldest parts that are removed. The critical basal parts of the plant, where leaf growth takes place and new leaves emerge, remain intact. Grasses are thus highly adapted to a regime of frequent, intense cropping, because they sacrifice the least functional parts of their food-producing machinery while hiding the regenerative portions of the plant.

Before humans invented mowing, the chief agents favoring this style of plant architecture were large grazing mammals—horses, cows, sheep, elephants, kangaroos, and the like—which for the past ten million years or so have inhabited all the world's major continents except Antarctica. Tough fibrous leaves and tiny bodies of silica—the primary mineral of which beach sand is made—enhance the defense of grasses by toughening the plant, wearing down the teeth of consumers, and therefore slowing the chewing process. The high-crowned molars of most grazers are well suited to this abrasive diet, for they grow continuously even as the grinding surfaces are worn down.

There are, of course, other ways for plants to thwart herbivores. Stems and leaves may bear spines that interfere with feeding; others are covered with hairs, which impede the progress of insects as they pierce or cut edible parts of the plant. Chemicals poison, deter, or repel herbivores of every description. Fibers and woody tissues resist physical or chemical breakdown in the digestive systems of would-be consumers. Still other defenses minimize the effects of inevitable damage. Multiple growing points ensure that a plant can continue to grow even if some of the tasty young shoots are eaten. The network of major and minor veins on the leaves of most flowering plants and some ferns allows water and sugars to move into and out of the leaf even when some of the conduits of transport are severed. These and other countermeasures to herbivory, either alone or in combination, work well under many conditions of exploitation. The key to understanding the modern success of grasses is thus to identify the situations in which the defenses and other attributes of grasses are superior to those of other plants, and to document how plant-eaters and the environments they live in have changed through the ages.

One long-held view is that grasses acquired their unique features and then became important components of the world's vegetation in response to climatic change. The great grass-dominated prairies, steppes, and savannas of the world thrive in regions where annual precipitation is less than about five hundred millimeters. Closed-canopy forests, where grasses constitute a small fraction of plant cover, grow under conditions of higher rainfall. As climatic conditions became increasingly favorable to grasses during the past twenty-five million years, so the argument goes, at least ten lineages of mammals gave rise to highly mobile grazers, which roamed the new dry-adapted grass ecosystems in herds beginning ten million years ago.

The argument against this explanation rests on two intriguing discoveries. Fossil deposits containing the silica bodies of grasses indicate that these plants originated in the understory of forested habitats, not in dry open areas. The oldest occurrences so far known are from the latest Cretaceous period, about seventy million years ago, in India; but related plant families may date back to one hundred million years ago or more. In a late stage of their history near the end of the Cretaceous, large herbivorous dinosaurs were living alongside small mammals. However, it was not until much later—thirty million years ago in South America, twenty-five million years ago in North America, and perhaps fifteen million years ago in Africa—that grasses became the signature architects of landscapes, which they have remained ever since. Soils characteristic of tall-grass and short-grass prairie vegetations became common in North America in the Great Plains about twenty million years ago, at a time when hoofed mammals were evolving rapidly. According to detailed studies of teeth and skulls of herbivorous mammals carried out by Christine

Janis, a brilliant vertebrate paleontologist at Brown University, these early mammals were either browsers, specialized on a diet of tree and shrub vegetation, or mixed feeders on a wide variety of plants, including grass. The high-crowned teeth of grass-eaters appeared in numerous lineages of mammals whose diet included large amounts of fibrous plant material as well as gritty dust.

The idea that herbivores were responsible for the evolution of the peculiarities of the grass habit, as well as for the subsequent development of grasslands, is supported by the second discovery. Although grasslands and the grazers inhabiting them occur on all the major continents, they are largely absent from the native ecosystems of Madagascar and New Zealand, places where the climate is conducive to grasslands and where such grasslands have become widespread since humans introduced domesticated grazers to these islands. Extinct moas—giant flightless birds, some weighing up to two hundred kilograms—and living flightless kaka parrots evolved as incidental grazers in New Zealand, but only in parts of the South Island above treeline did grasses form the dominant component of the vegetation. No other herbivores on these or other smaller land masses developed the appetites associated with grass-feeding in continental mammals. In other words, a suitable climate is not sufficient to explain the success of grass.

This observation implies that grasslands thrive only in places where herbivores are large enough and common enough to restrict trees and shrubs, which might dominate the same environment in the absence of grazers. In other words, the intensity of plant consumption in grasslands is so high that the defenses of other kinds of plants are no longer adequate. Only plants whose vulnerable growth zones are hidden from the depredations of plant-eaters can flourish as dominant members of the vegetation in the face of intense grazing.

Why, then, would the huge plant-eating dinosaurs of the Jurassic, weighing up to seventy tons, not have selected for grasslike defenses in plants of that period? One possibility is that the long necks of many of these dinosaurs enabled the animals to feed high in the trees. The problem with this hypothesis is that pumping blood to the head with the neck raised requires a much larger heart and higher blood pressure than these dinosaurs likely had. As in giraffes, the long neck of dinosaurs may chiefly have been associated with male-male combat. Long-necked giant dinosaurs may indeed have been able to feed on twigs and leaves up to ten meters (about thirty feet) above the ground, but most of their food probably grew as low vegetation.

The only answer that makes sense to me is that, despite their enormous size, these ancient herbivores ate less fodder and consequently imposed a less intense regime of consumption on the vegetation than do modern grazing mammals, the largest of which reach a mass of ten tons (African bull elephants). Dinosaurs and other large plant-eaters living before the Late Cretaceous must have had lower levels of activity, and therefore a lower demand for food, than do modern large mammals. They could not have had the warm-blooded physiology that characterizes living mammals and birds, whose food requirements are four to ten times greater than those of equivalent-sized "cold-blooded" lizards and turtles. Grasslike defenses would, of course, have been useful even under a more relaxed regime of herbivory, but under such a regime, faster-growing trees, shrubs, ferns, and horsetails with other defenses might have been able to hold a competitive edge. Indeed, the dominant plants of Jurassic vegetations—conifers like *Ginkgo* and *Araucaria,* understory horsetails and cycads, and ferns—would have formed the bulk of dinosaur diets.

The idea that the food intake of the great Jurassic dinosaurs was modest is supported by inferences about the digestive systems and growth rates of these animals. Recent studies of the growth of the long bones of the leg indicate that Jurassic herbivorous dinosaurs grew two to five times faster than reptiles of comparable extrapolated size, but five to six times slower than warm-blooded mammals and birds. Instead of breaking up fibrous plant material by grinding it in the mouth with cheek teeth, as mammals do, the large Jurassic plant-eating dinosaurs fed more like birds and turtles, by plucking plant parts with peglike front teeth and then swallowing large pieces whole. Slow digestion, likely aided by bacteria, as in most other large herbivores, took place in the cavernous gut. Fully digesting large chunks of vegetation takes longer than digesting small fragments, because each piece presents a relatively small area for the large mass that digestive enzymes must break down chemically. Small fragments, by contrast, have a large surface area per unit mass. Everything we can infer about the physiology of giant dinosaurs points to relatively low food requirements and lower metabolic rates in these animals than in equivalent herbivores of today.

If the dominant herbivores of the Jurassic period were ineffective enough to prevent grasslike plants from gaining an edge over plants whose growing points are near the tips of branches, why did chewing herbivores with bigger appetites not displace them? Why did vertebrates who chew their food rise to prominence only in the Late Cretaceous, and why did grasslands become established no earlier than thirty-five million years after the close of that period?

It is tempting to answer such questions within the narrow context of plants and their consumers on land. In that approach, the solutions tend to refer to the particulars—the physiology of dinosaurs, the characteristics of plants, the climatic regimes in which the consumer and the consumed execute their evolutionary dance—but such a tack brings with it the risk that other relevant facts not directly pertaining to the interaction under scrutiny are overlooked. By taking the broader view, we can often identify causes and explanations that would have remained elusive had we confined our analysis to the specifics of the situation.

The first thing to note is that the grasslike habit is a late innovation not just among land plants, but also in the sea. Sea grasses—flowering plants distantly related to true grasses but with a similar basal growth zone—entered the sea during Late Cretaceous time. They have since become immensely important in coastal environments worldwide. Sea-grass meadows are among the most productive marine environments on Earth and serve as habitats for countless species. Another group of marine plants—the kelps—have also evolved the grasslike habit of growth. The ends of the blades of these very large brown seaweeds are the oldest parts of the plant, whereas new tissues are formed near the base of the blades. The bulk of the evidence available implies that kelps arose only some thirty-five to thirty million years ago in the North Pacific, spreading from there to other cold-water environments. Like sea-grass meadows, kelp forests like those on the coasts of California, Chile, New Zealand, Nova Scotia, and South Africa are immensely productive habitats. Other seaweeds, like most land plants other than grasses, grow at the tips of branches or along the edges of fronds. The late arrival of the grasslike habit is therefore not only a phenomenon of the dry land, but one that extended to the marine realm as well.

If herbivores are responsible for the success of grasses on land, might the same be true for plants with a similar growth habit in the sea? To answer this question, I needed to know about marine plant-eaters. How do they consume plants? When and how did they evolve? Did the origins of sea grasses and kelps coincide with the evolution of large herbivores with high food requirements?

I had been collecting evidence about such matters for some time when I brought up the subject with my Berkeley colleague David Lindberg. By coincidence, Lindberg—an expert on plant-eating cap-shaped limpet snails—had also become interested in the evolution of marine herbivores, and had reached the same preliminary conclusion: animals that consume plants in the sea are a relatively recent addition to the evolutionary tree of life. Joining forces, we set about collecting more data, and soon had a preliminary draft of a paper. The evidence was clear enough, confirming our initial impressions; but we found the results so counterintuitive that we held off submitting the paper for more than a year. Our intuition had told us that, with plants at the bottom of the food chain and predators at the top, we would expect plants to evolve before herbivores, which would evolve before their predators. But instead we were now confronted with several lines of evidence that herbivores were relative latecomers, and that predators, rock-scrapers, and scavengers had appeared long before the habit of feeding on living marine plants evolved. Either our intuition was wrong, or our data were seriously flawed or misleading. In the end, we decided that it was our initial expectation that was misplaced.

When we arrayed all the animals capable of ingesting seaweeds and sea grasses on the evolutionary tree, we saw at once that all were concentrated at the tips of the branches. Ancestors that scraped rocks or fed on decaying matter repeatedly gave rise to herbivorous lineages of snails, worms, crabs, sand hoppers, sea urchins, and even a few sea stars; whereas plant-eating lineages rarely produced animals with such specialized food habits as predation, filter-feeding, or scavenging. Furthermore, very few plant-eating lineages have a fossil record older than about two hundred fifty million years ago, the time marking the beginning of the Triassic period of the Mesozoic era. Herbivorous sea urchins and cap-shaped limpet snails belong to lineages with Late Triassic origins, even though the larger evolutionary branches to which they belong go back hundreds of millions of years earlier. The earliest plant-eating fish are known from the Early Eocene epoch, about fifty million years ago, and although the fossil record of fish is spotty enough that an earlier origin—perhaps in the Late Cretaceous—cannot be excluded, all earlier fish, dating back to some four hundred forty million years ago, were either predators or filter-feeders. Plant-eating marine vertebrates with voracious appetites—green turtles, geese, sea cows, and a bizarre lineage of Late Miocene to Early Pliocene sloths from Peru—have Late Cretaceous or Cenozoic origins. The vertebrates are and were major consumers of seaweeds as well as sea grasses. Just as on land, therefore, large herbivores with big appetites may have given the decisive advantage to marine plants whose vulnerable growing parts are concealed from relentless exploitation.

Seaweeds, on the other hand, are known to date back to at least 1.2 billion years ago, some six hundred million years before the rise of multicellular animals. Early seaweeds were mere filaments and sheets, but rather complex forms with branches had evolved by the time of the earliest animals. Large-bladed seaweeds, including the basally growing kelps, seem to have evolved much later in several of the major groups of seaweeds.

What we know so far is that plants with a grasslike habit and the animals that consume them evolved late, and at about the same time, both on land and in the sea. Why did these events not take place much earlier? Was some sort of global constraint, which was in place for hundreds of millions of years, finally lifted during Late Cretaceous time? If so, what was this constraint, and what caused it to be overcome?

The possibility that a lid on productivity was lifted deserves special attention. Rapid, sustained consumption by large herbivores can be maintained only when food-producing vegetation can keep up with demand. The most compelling evidence that plant productivity on land began to rise as the metabolic rates of top consumers increased during the Late Cretaceous comes from recent work on fossil leaves by Kevin Boyce, a leading paleobotanist at the University of Chicago, and his colleagues. In photosynthesis—the process by which plants use light, nutrients, and carbon dioxide to make living tissues and release oxygen—plants trade water for carbon. The more water a plant transpires into the atmosphere through pores in its leaves, the more carbon dioxide it can take up through those same pores, and the more biomass the plant can make. Water is transported within the leaf from the stalk to the mesophyll—the tissue where the machinery of photosynthesis is situated—by a system of branching veins, which are most clearly seen on the leaf's underside. If the density of veins is high, water can be brought close to every site of photosynthesis, making it possible for the leaf to transpire water and to make biomass quickly.

By measuring the vein densities in fossil leaves, Boyce and his team discovered that all the land plants living before about one hundred million years ago had uniformly low vein densities, averaging about two millimeters of vein length for each square millimeter of leaf surface, and rarely exceeded five millimeters of vein length per square millimeter. All this began to change when flowering plants, which began as small weeds during the Early Cretaceous about 140 million years ago, started to diversify in mid-Cretaceous time. Flowering plants, especially those belonging to the so-called eudicot group, have average vein densities of eight millimeters vein length per square millimeter of leaf surface, four times higher than in their predecessors. Some have as much as twenty-five millimeters of vein length per square millimeter of leaf. These differences correspond to a three- to fourfold increase in the rate of transpiration and biomass production beginning about one hundred million years ago. The low rate of photosynthesis per leaf in ancient plants could have been compensated for by a larger number of leaves per plant, but Boyce and his team argue that the architecture of fossil plants makes this unlikely. The number of leaves on a given area of ground was, if anything, lower before the Late Cretaceous than afterward, indicating that the whole-plant rate of photosynthesis rose substantially as flowering plants took over the land in most parts of the world.

The fascinating implication of this finding is that the pace of life—the rates of production, consumption, and interaction—must have quickened. If so, we should expect other manifestations of a high-energy way of life to have evolved at about the same time flowering plants with high vein densities, plants with continuous growth at the base of blades, and the animals consuming these plants expanded.

This is indeed the case. Rapidly metabolizing warm-blooded mammals, birds, and perhaps some advanced predatory dinosaurs came into their own during the Late Cretaceous, along with highly metabolically active social insects, including bees, wasps, early ants, and termites. Marine predators capable of maintaining high core body temperatures first evolved in the sea during the Cretaceous (some lineages of sharks) and the Eocene (tunas and related fish), while other warm-blooded birds and mammals entered the sea to become demanding consumers. This more frenzied existence, driven by intense competition at all levels in the food chain for essential foodstuffs, is contingent on a high and predictable supply of raw materials for the plants, as well as on enough oxygen to power high rates of activity of top consumers. These essentials, in turn, are under the control both of physical processes on the Earth—volcanism, erosion, and the production of new crust—and of the collective physiology of life itself.

Huge volcanic eruptions, which poured millions of cubic kilometers of magma on to the ocean floor during episodes throughout the Cretaceous period, released enormous amounts of carbon dioxide, which promoted plant growth, as well as minerals like iron that are essential to the growth of marine plankton. Repeatedly during the Cretaceous, vast quantities of organic matter were buried in sediments beneath the oceans during so-called oceanic anoxic events. This excess buried carbon did not combine with oxygen and was therefore effectively removed from the Earth's natural economy, leaving the ocean and atmosphere richer in free oxygen. Geological processes may therefore have acted as triggers, but organisms were responsible for capturing and effectively deploying the resources that had been made available by these processes. Capitalizing on the geological bonanza, living things accumulated and recycled nutrients to create an ecological and evolutionary feedback loop in which high-energy ways of life became possible and sustainable on a worldwide scale.

Webs of interdependence extend far beyond herbivores and their food plants. Plant roots, especially those of trees, mechanically break up rock particles in the soil, and secrete acids so that the minerals that become exposed on the newly created surfaces of the resulting fragments are dissolved and made available to the plants. Microbes and fungi, which form intimate partnerships with the plant roots, aid in this mining operation. At the other end of the food chain, predators impose intense selection on herbivores and keep the latter's populations in check. Without highly active, hungry predators, there would be less evolutionary incentive for herbivores to reach large size—and thus a certain immunity from attack—and there would be less incentive for plants to conceal their zones of growth from herbivores. Cooperative arrangements between soil microbes and plants would also be less strongly favored. All the species in these complex webs of interaction are therefore adapted to, and have collectively created, an environment of high productivity in which resources are either themselves organisms or are under the control of organisms.

A particularly striking example of this is the cultivation of one species by another. Like specialized marine herbivory and the interdependence between grasses and grazers, natural agriculture is known only from the last one hundred million years or so of Earth history. For example, leaf-cutter ants in tropical America collect leaves from living plants and bring them to underground chambers, where the leaves serve as food for domesticated fungi, which are nurtured, weeded, kept pest-free, and consumed by the ants. Fungal cultivation may have begun as early as fifty million years ago, but its most specialized manifestations are no older than ten million years. Another example is a tropical western Pacific reef-dwelling damselfish that tends gardens of a seaweed, which occurs only in those gardens. Damselfish originated some time in the Eocene epoch, not more than fifty-five million years ago, but it is likely that the specialized aquaculture practiced by the species in question is much more recent. In these two cases, and probably in other examples of agriculture, the species being cultivated has been domesticated—that is, it has been exposed to a pattern of selection by its host that differs from the selection to which undomesticated relatives are subjected. Hosts have generally selected for higher yield, easier handling, and disease resistance, and likely reduced toxicity and other defenses the cultivated species might have had before domestication.

In nature, such cultivation and domestication are limited to insects (for example, ants, termites, and bark beetles), crustaceans (mud shrimps cultivating bacteria on pieces of sea grass deep in tunnels and galleries beneath tidal flats), and snails (some gardening shore limpets). All the known hosts eat plants, microbes, fungi, or their products; none is a predator. Some ants tend honey-producing aphids and bugs, but they eat the honey and not the insects that produce it.

The case of domestication by humans stands apart from other evolutionary acquisitions of agriculture for three reasons. First, we are alone among tetrapods—mammals, birds, reptiles, and amphibians—in domesticating other species. Second, we include among domesticated species not only plants for food and fiber, but also animals for meat, milk, transport, and labor. Third, domestication is a remarkably recent achievement in the history of our species, one that was attained independently in perhaps as many as twenty-four societies. Domestication occurred in environments ranging from seasonally dry grasslands to rain forests in parts of the world as varied as southwest Asia, southern India, China, New Guinea, northern tropical Africa, southern Mexico and Central America, eastern

North America, and the mountains and lowlands of South America. Our species endured the ice ages in Europe and Asia, and prolonged droughts in Africa, for tens of thousands of years before making the transition from hunting and gathering wild foods to cultivating and then domesticating some of these species beginning as early as thirteen thousand years ago in the Middle East. All pre-modern geographic expansions of humans took place before newcomers invented agriculture. Northern Europe was occupied by humans for at least thirty-five thousand years before agriculture was brought there from Asia, and the Americas were colonized by people from Asia three to four thousand years before several societies cultivated and domesticated crops in tropical America.

It is no exaggeration to say that our species owes much of its modern world dominance to herding and agriculture. By increasing the yield of crops and stabilizing the food supply, the human population increased by a factor of one hundred or more. Agriculture set the stage for urbanization beginning some five thousand years ago by producing so much food that new ways of life not directly tied to food production became possible, including occupations in government, the arts, education, trading, and the clergy.

The late invention of agriculture by humans poses a puzzle. Why would humans not have begun to domesticate food organisms much earlier if our cognitive skills had already reached modern levels of sophistication perhaps as early as one hundred thousand years ago in Africa, as indicated by the discovery of ornaments in South African and Mediterranean archeological sites? One possible answer is that the full complement of human intellectual capacities was not achieved until thirteen thousand years ago, but I find this unlikely. If that explanation were correct, it would mean that cognition improved in parallel to modern levels in twenty-four separate societies worldwide. A more probable explanation, it seems to me, is that environments worldwide became productive enough after the waning of the Pleistocene ice ages to support domestication and the accompanying human population expansion. The benefits of agriculture would presumably always have existed, but its feasibility may not have. Once agriculture had taken root under the more productive conditions of post-ice-age environments, it could spread even to less favorable places, such as those requiring irrigation or deliberate fertilization of the soil. The trigger, then, may have been the end of the ice ages. In many parts of the world, including the Middle East, this ushered in a warmer, moister climate with more rainfall for plant growth. If this hypothesis proves to be correct, it would mean that the interdependency represented by human agriculture is like that seen among producers and consumers elsewhere in nature in being made possible by an increasing productive potential of the environment.

Interdependence may provide added value and economic benefits to all members of an economy, whether in the natural world or in the human domain. But it also exposes an economy to catastrophic collapse when essential links are severed. This is most obvious when the means of production are disrupted, such as when land plants and marine phytoplankton are deprived of sunlight for extended periods or when the usual pathways of supply of such essential raw ingredients as water, nitrogen, and phosphorus are cut off. A bottom-up collapse of this kind, triggered by such calamities as a collision of the Earth with large objects from outer space or a gigantic volcanic outburst that releases dust and poisonous gases into the atmosphere is likely the cause of history's great mass extinctions. However, it is amplified by the great dying itself, which puts an end to normal cycling of nutrients and which, through decay on a gargantuan scale, may diminish the amount of free oxygen.

Human history, too, witnessed convulsions stemming from long droughts, volcanically induced crop failures, and long cold spells. In China, for example, periods of dynastic overturn have been attributed to climatic events that interfered with agricultural production.

In all these natural and societal collapses, the pressure of consumption by surviving members of the system under conditions of low resource supply amplify the initial shock as the intricate interdependencies that have developed during more favorable times become unraveled.

We have also learned, however, that the patterns of mutual dependence that have built up over evolutionary time are severely compromised when top predators are eliminated from a system. The regional extinction of gray wolves in the western United States released controls on populations of herbivorous deer and elk, causing massive intensification of herbivory on the vegetation. Decimation of sea otter populations in the North Pacific enabled sea urchins and abalone—favorite prey of the sea otter—to multiply, in turn decimating stands of leafy seaweeds and giving encrusting rocklike coralline seaweeds the opportunity to form "barrens" over wide areas. Removal of active herbivores like bison, elephants, and sea cows has the paradoxical effect of decreasing the productivity of the plants consumed by these herbivores, and therefore giving other plants opportunities. Chronic overfishing on reefs has diminished densities of predatory and herbivorous fish, turtles, and marine mammals by 90 to 95 percent, greatly increasing the entire reef ecosystem's vulnerability to disease and storm damage. In such highly exploited ecosystems, the control that consumers collectively exercise is removed, placing producers at the mercy of outside forces.

The destructive consequences of overexploiting top consumers are already being felt in the coastal zones of most of the world's oceans. Since the middle of the twentieth century, there has been an exponential rise in the number and size of so-called dead zones, areas of ocean floor where bottom waters and sediments are devoid of oxygen. The principal cause of the expansion of dead zones is the huge influx of nutrients, especially nitrogen and phosphorus, in fertilizers and pollution from farms and cities on land. The amount of nitrogen reaching the ocean today is approximately double what it was before artificial fertilizers came on the market. Nutrient enrichment of the coastal zones has coincided with a worldwide decimation of fish and marine mammals, especially including top-level predators—sharks, rays, tunas, cod, grouper, billfish, whales, and seals—but also grazers (sea cows, sea turtles, and various reef fish) and suspensions-feeders like herring, sardine, anchovy, and clams. Suspension-feeding oysters have been all but eliminated from estuaries all over the world. Nutrients that would normally have been channeled into and recycled by these consumers are now passing directly to microbes, which are not being consumed and which take up most of the oxygen. This problem is especially severe in fertile areas where surface and bottom waters fail to mix. With an intact community of consumers, marine ecosystems in nutrient-enriched sectors would likely be far more resistant to enrichment, and would not face the real threat of collapse.

Organisms do not live in a vacuum. They are independent actors, pursuing their self-interest as competitors, but they are also linked to one another in a complex web of relationships. They create a civilization of sorts, a system of antagonisms and productive partnerships in which participants are well adapted to each other, a whole that nourishes life and that thrives on adaptation. If that civilization is disturbed, its constituents will suffer. Humans are part of that civilization of life. We benefit from it and we cannot escape its embrace. Given our power and ambition to alter the world, we have a special global responsibility to do what we can to promote that civilization.

Even though the ultimate explanation for the success of the grass habit remains speculative, the facts in the case persuasively show that preconceptions and intuitions are poor guides for discovery. Without an appreciation of history, we would never have known that such a commonplace fixture as grass, like the suburbs in which it flourishes so well in the human domain, is a relatively recent addition to the world's flora. Without fossils, a well-supported evolutionary tree, and knowledge of the natural history of feeding, we could not have learned that high-energy herbivores are responsible for the global success of grasslike plants, and that such plant-eaters are themselves latecomers. Without the discipline of scientific inquiry, grass is just grass—unremarkable, pleasant enough, even worth a poem or two—a minor ornament in our lives. No longer can we take grass for granted, or grant it an automatic place in the iconography of a dinosaur's world.

CHAPTER EIGHT

Nature's Housing Market, or Why Nothing Happens in Isolation

It has become a tradition that, when a doctoral candidate is subjected to an oral examination by a committee of professors, someone on the prosecution side asks a trick question. In the examinations I attended, such a question often comes in the form of a puzzling specimen. The aim is to see how much the student can infer about the composition, identity, and life habits of the animal or plant that is represented by the object under scrutiny.

Sometimes, however, the tables turn, and the trick is on the examiner. One case involving hidden identity and new insight into the nature of adaptation stands out in my memory. As part of a successful oral examination of a doctoral candidate in population biology at Davis, one of my fellow interrogators began a line of questioning by passing around a slender, slightly curved tube, open at both ends, with a small mushroom-shaped object attached at the narrower end. My colleague wanted to know how many phyla—major evolutionary branches of animal life—were represented by this specimen.

To most people in the room, the answer seemed almost too easy. "Two," the student answered confidently. The tube was clearly a tusk shell, belonging to the class Scaphopoda in the phylum Mollusca;

and the mushroom-shaped object was a cup coral, in the class Anthozoa of the phylum Cnidaria. To my surprise, the inquisitor agreed. All the while, I had been convinced that he laid a trap for the student, a quite ingenious trap because of the phantom presence of at least one additional phylum. I thus interrupted the proceedings to pursue the matter further.

I invited the group to examine the specimen more closely. The tube's outer surface was pitted and in places obscured by tiny growths, a condition indicating that the shell had lain on the sea bottom for some time. In life, scaphopods are buried in sand or mud, where organisms that corrode or encrust the shell surface cannot settle. In other words, the scaphopod was dead when its shell was collected. Next, I drew attention to the coral. There was, I explained, a specific and consistent association between this kind of solitary coral and a particular kind of worm—a sipunculan worm—that always inhabits discarded shells, including those of scaphopods. The worm leads a mostly sedentary life on sandy and muddy bottoms well below the low tide line in the tropical Pacific and Indian Oceans. The specimen thus contained three phyla, not counting the creatures that pitted and encrusted the tube's outer surface. The student was wrong, but so was the initial examiner. We all learned something, and the exam ended with a unanimous and enthusiastic vote to pass the candidate.

Shell-dwelling sipunculans may be unfamiliar to most people, but the habit of living in discarded shells is common, with members of nearly every major animal group taking advantage of the prolific and safe environments that shells provide. Best known among secondary shell tenants are hermit crabs, crustaceans whose soft abdomen fits neatly into the interior cavity of an otherwise empty shell. Unlike lethargic sipunculans, hermit crabs typically move about actively, carrying their acquired house with them. Some small fish and octopuses also live or brood in shells, which they guard against potential rivals. Trilobites dating from almost five hundred million years ago have been found fossilized inside the shells of ancient cephalopods, indicating that shells have served as secondary dwellings for mobile animals for almost as long as there have been shells.

Inevitably, other animals have become specialized as parasites of these secondary shell dwellers, and still others live sedentary lives attached either to the outside or inside of shells. Some segmented worms, for example, build tubes inside shells inhabited by hermit crabs, and either steal food from the host or consume some of the host's eggs. In most parts of the world, there are small cap-shaped limpet snails that spend their lives on the outsides of shells, in which some of them excavate pits so that the limpets cannot be easily dislodged. Life on a mobile substrate offers protection from many predators such as fish, snails, sea stars, and crabs, which cannot or will not venture onto a small moving surface to look for prey. In other words, shells are microcosms—miniature communities in which the universal economic relationships of competition and cooperation are played out in all their complexity.

Intriguing natural history and fodder for oral examinations aside, housing offers a window into how the economy of life works. The shell, originally built by a mollusc such as a snail, scaphopod, clam, or cephalopod, is a highly adapted part of a working creature living and propagating in a competitive world. Although its primary function is protection from predators and inclement conditions, the shell has, in some lineages, also become a weapon during aggression, a site for laying eggs, a ballast tank (in shell-bearing cephalopods like *Nautilus*), an aid to locomotion, and a display organ. As an integral part of the mollusc, the shell reveals a great deal about the habits and conditions of life of its maker, especially the dangers and opportunities that collectively constitute the regime of selection under which the animal evolved and still thrives. When empty, the shell becomes part of the housing market for secondary shell dwellers. It is a commodity, a resource for which tenants compete.

Adaptation and economics are thus inextricably linked. Success in survival and propagation depend on whether living things have access to sufficient resources and whether they can defend those resources when challenged. This is as true in our economy, from the local real-estate market to the global marketplace, as it is in the economy of nature. By thinking about evolution in economic terms, we discern close parallels between our economic decisions, whether they be rational or irrational, and the ways in which natural selection has shaped the allocation of resources among evolving life-forms. Housing exposes these parallels with clarity, and illuminates the feedbacks that bind consumers and resources together to create an emergent ecology. It is no accident that the words *economy* and *ecology* are derived from the same Greek root *oikos,* meaning "house."

A glance at our own dwellings tells us a great deal about the economic and social status of their occupants and about the conditions that those occupants deem important. We instantly recognize wealth and status when we enter the grounds of a grand mansion. Secure gates and barred windows speak of worries about crime, as do the moats and thick walls surrounding a medieval castle. On Guam, an island periodically ravaged by monster typhoons and by earthquakes registering eight on the Richter scale, houses are built of massive concrete, with flat roofs; an ovenlike design that is jarringly at odds with the hot, sunny climate; and one that would be unlivable without air-conditioning with cheap electricity. In Holland, well-situated houses have large windows facing east and west to take advantage of the warm sun in the cold, damp winter. In many parts of the wet tropics, houses are made of bamboo poles situated well above the ground. They are open structures designed to catch cooling breezes and to keep snakes out. In short, houses reflect our stations in life, much as snail shells reflect the life of the original builder and the life of later tenants. Their architecture is a material expression of how we, and other house-builders, respond to and construct the physical and social environment.

Of course, the same is true for the shells, tubes, and nests built as protection by other animals. An abundance of predators that break, enter, or drill their way into shells and tubes in shallow tropical seas enforces a high standard of sturdy construction. For molluscs that move too slowly to get away from predators, the shell must have a thick wall, a small opening with a reinforced rim, and some device like a door that allows the vital organs inside the shell to be hermetically sealed from outside danger. These requirements are relaxed in situations where breaking and entering is less likely, as in the deep sea, in polar regions, and on the bodies of corals or toxic plants where predators are loath to search for prey. Where resources are scarce, houses are constrained to be small. Only the best competitors and the most predation-resistant species can afford to become large animals with palatial shells.

Once the original molluscan builder has died and the shell is taken over by a secondary occupant, the shell becomes part of an animal with a completely different evolutionary past. The new occupant usually lives under circumstances that in important ways differ from those in which the shell's adaptations evolved and in which its builder prospered. Though not evolutionarily "designed" as part of its new occupant, the shell still functions adequately as part of that animal. Despite obvious shortcomings and its tendency to deteriorate under occupation by a resident who is incapable of making repairs, the shell and its adaptations continue to work well enough for a succession of hard-living renters to benefit from occupying it.

Nowhere is this more obvious than in hermit crabs that live in clam shells. Although most hermit crabs live in the shells of snails or scaphopods, a few are specialized for life in one half of a clam's two-part shell. Living clams have shells consisting of two valves joined at the top by a flexible ligament. One or two muscles stretch from the inner side of one valve to the inner side of the opposite valve. When these muscle contract, the valves shut; when they relax, the ligament acts as a spring to spread the valves apart, enabling the clam to extend and deploy organs for feeding, respiration, and locomotion. Once the clam dies, the two parts of the shell usually separate from each other. Many of the shell features that made up part of the living clam's adaptive package become irrelevant for a hermit crab that hides its abdomen in the broad hollow of one of the discarded valves. External ribs that stabilize a buried bivalve's position in the sand during life no longer have this function for a hermit crab, which moves about on the sea bottom and carries the valve over its body. And yet the valve, built by an entirely unrelated species in a radically different habitat, is still part of the adaptive package of a species for which that shell was not designed. The shell functions well for both animals; its characteristics collectively are an adequate hypothesis of the evolutionary environments of two species whose last common ancestor lived as much as 580 million years ago, well before the two lineages they represent separately evolved a hard skeleton.

The co-option of one organism's adapted structures by another is by no means confined to secondary shell dwellers. Weaver ants construct communal shelters by folding and connecting living leaves with silk. Birds may line their nests with the fur of mammals, and bowerbirds ornament the sites where they seek to attract females with snail shells, beetle-wing covers, and other artifacts built by animals and even by humans. Many sea slugs arm their vulnerable external gills with the unexploded nematocysts (stinging cells that release their contents upon contact with a predator or food organism) of the sea anemones or hydroids on which they prey. In these and hundreds of other cases, structures that fulfill functions in their original bearers become part of the adapted complex of another.

Like human renters, most secondary shell dwellers do not remodel or repair their houses. Attacks by predators, infiltration of the shell wall by borers, and the wear-and-tear of being dragged over rough ground take their toll on the shell's worthiness as a dwelling. Again, like apartment dwellers with little stake in their temporary quarters, most hermit crabs frequently move house. Intense competition in the housing market often involving aggression and even eviction reflects a locally limited resource of good housing, and leads to a redistribution of that resource according to who has high status and who must make do with hovels. Given an array of shells among which to choose, hermit crabs typically select a larger abode than the one they occupy, one with sufficient room inside to accommodate the abdomen and (for females) the eggs attached to the abdomen, and with sufficient protection to ward off predators. Studies show that most members of a hermit-crab population live in substandard housing—damaged, overgrown, weakened by borers, and too cramped—and yet, despite its inadequacies and imperfections, the housing is good enough. It suffices even if its occupants, like most people, would do better in a bigger place.

The important point that emerges from this natural history is that adaptations, even if they arose under a selective regime that is quite specific to one species, are beneficial under a wide range of conditions. Molluscan shells possess adaptations that are essential to the survival of their makers. Temporary tenants have co-opted these adaptations for their own ends. The adaptations that these tenants have thus acquired are adequate—indeed indispensable—in the new context in which the shells now function. Adaptations are often surprisingly versatile, regardless of their particular circumstances of origin.

Shells that become newly available to a population of hermit crabs do not remain empty for long. It is ironic that the supply of housing is not only controlled, but stabilized and enhanced, by predators, the very animals against which shells offer protection. Many predatory snails, sea stars, flatworms, and fish leave shells of their prey essentially intact, and therefore produce a steady supply of fresh empty shells year-round. Some hermit crabs have learned to wait near sites where such predators feed, so that they are first in line for a pristine dwelling. In the absence of this predictable source of supply, suitable new shells would become available only during episodes of mass mortality of snails caused by excessive heat, cold, or stormy weather. These episodes would also kill the potential new occupants and thus reduce the demand for housing. Huge numbers of shells would flood the market at unpredictable times, only to disappear unused. At other times, almost no shells would come on the market, so that demand vastly outstrips supply. Under such a boom-and-bust regime of supply, the shell resource cannot be effectively exploited. Only when that supply becomes predictable, as it does when organisms—predators, in this case—control it, can the resource become a target of exploitation and adaptation. Supply and demand are thus evolutionarily linked through an unintentionally evolved, collective form of resource regulation, in which living things that compete among themselves nevertheless work together to form complex networks of interdependence.

It is interesting that few, if any, hermit crabs commandeer housing directly from snails. If they did, they would kill the goose that laid the golden egg, for the supply of shells might be jeopardized, rendering the shell-dwelling habits unsustainable. Nature's way of regulating supply is thus like that in our own economy: consumers do not control their own resources, but third parties do. In the case of the shell resource, it is predators that control the rate of production as well as the quality of housing. Regulation acts through feedbacks among several players in the economic web; it is circuitous and indirect, and because of that diffuse control, the resource is exploited sustainably.

I don't wish to leave the impression that hermit crabs are entirely beholden to their molluscan architects for adequate defense. When they retreat into the shell, hermit crabs of many species can close off the shell opening with one of their front claws, which thus acts as a door. And these doors can bite. My fingers have often been surprised by a powerful nip as I ascertained which species had made its home in a particular shell. When such aggressive behavior is accompanied by a growling sound emanating from within, as in the large hermit crab *Aniculus aniculus* from the tropical Pacific, one is left in no doubt that the defensive traits of the shell have been substantially enhanced by the occupant's own measures.

Moreover, not all secondary tenants are irresponsible renters. In deep waters and in other situations where the supply of suitable shells is low, many tenants have in effect hired an in-house—or, more accurately, an on-house—architect to improve or even to enlarge the dwelling. Sea anemones, sponges, moss animals (bryozoans), and a variety of other sedentary creatures often settle on the shell of a hermit crab or sipunculan worm. For the occupant inside, these external encrusters often provide additional protection. The cup coral at the narrow end of the scaphopod shell inhabited by a sipunculan worm, for example, likely defends its host against predators by deploying its stinging cells (nematocysts). In still more intimate partnerships, the encruster surrounds the entire shell and eventually enlarges the space for the occupant within by building an extension, eliminating the necessity for the crab to move to roomier quarters as it grows.

I became acquainted with a truly extreme version of such a modified house during one of my many visits to Guam. As part of a study on the feasibility of commercially trawling for deep-sea shrimp, biologists in Guam had brought up hermit crabs living in a peculiar, flexible shell completely encased by a sea anemone. The shell, golden in color and of a somewhat sloppy spiral form, bore growth lines, and certainly could be mistaken for a somewhat nondescript snail snail, especially if the anemone's flesh were stripped away from its exterior. However, no one had ever seen such shells with a living snail inside, and the papery consistency of the shell was all wrong for a mollusc. Instead of being composed of calcium carbonate, as is the case with most molluscs, it was made of chitin, a substance more usually associated with the skeletons of crustaceans, insects, and other arthropods. In 1895, shells of this type had come to the attention of William H. Dall, a world-renowned curator of fossil and living molluscs at the Smithsonian's National Museum of Natural History in Washington, D.C. Dall described the peculiar plastic shells as belonging to a new genus and species, *Stylobates aeneus,* which he tentatively regarded as a snail, or gastropod mollusc.

But appearances can be deceiving, and Dall changed his mind some twenty-five years later, when he suggested that the shell was in fact part of the sea anemone. His suspicions were finally confirmed by Daphne Dunn and her colleagues in 1981. Dunn, a world authority on sea anemones who was then working at the California Academy of Sciences, recognized *Stylobates* as the shell-like base by which the sea anemone attaches itself to an object. After the anemone settles on a tiny snail shell inhabited by a hermit crab, it quickly grows to cover the shell, and then proceeds to grow in the same direction as the shell would have, producing a spiral extension around the growing hermit crab inside. The anemone and the crab thus form an inseparable partnership, a mutually beneficial arrangement in which the anemone has a place to live in an environment where suitable hard surfaces are few and far between, and in which the crab acquires a permanent dwelling that is not only maintained and enlarged by the anemone, but also shielded from potential predators by the anemone's nematocysts. Whether the two partners benefit each other in food-gathering is unknown, but this seems a likely extra bonus of the arrangement.

Here, then, is a case where a resource and its user have become tightly linked. It is not even clear who is the host and who is the guest. Both the hermit crab—always a member of the genus *Parapagurus*—

and the *Stylobates* anemone have worked together to grow and maintain their own resource. Such an arrangement resembles the partnership between humans and their domesticated plants and animals, which resulted in a large increase and stabilization of the human food supply and also substantially changed human living patterns from a nomadic to a more sedentary, territorial existence. In both the human and the anemone-crab relationships, a new emergent whole is fashioned by the working together of species with widely different ancestries. The self-interests of each member in the partnership no longer diverge; instead, they coincide, together creating a common interest in favor of the mutualistic relationship.

In the economies of life, the emergent wholes must function adequately throughout their lives, from birth through development to maturity. This criterion of adequacy also applies to the components. The properties and actions of the whole thus affect not only the fate of the whole, but also the fates of its parts. If the whole dies, so do its cells; and if the cells don't work, the whole too is compromised. This is what happens when cancerous cells divide uncontrollably. The self-interests of dividing cells, or of their energy-producing mitochondria within cells, must be checked in favor of the body in which those cells and those mitochondria are embedded. Wholes exist because antagonisms among their parts have been tamed and channeled to serve the survival and propagation of those larger, more complex units.

Some shells would make excellent shelters if only a hermit crab could fit into them. Many tropical shells, for example, have a narrow, slitlike opening and a correspondingly narrow interior. Cowries, cone shells, and other shells of this form offer effective protection to any occupant that can withdraw fully into the interior. To fit into such a shell, however, a hermit crab must have a flattened body. Many tropical species indeed have such a compressed form and are specialized to live in narrow-apertured houses. Others have the usual body with a round cross section, but when they live for a time in a cone shell or cowrie, the body gradually takes on a flattened configuration.

These oddities of the natural history of animal housing have nudged me to think about adaptation in ways that depart from the conventional view of this all-important evolutionary process. In the traditional conception, a good fit between a living thing and its surroundings results from a pattern of natural selection in which predators, competitors, disease agents, or even bad weather sort vulnerable individuals from those who, by virtue of possessing particular genetically heritable traits, survive to leave more offspring in the next generation. The would-be adaptations conferring this advantage were first tested in a particular set of circumstances, but their presence often benefits individuals under many other conditions as well. Abundant experimental evidence shows that this environment-driven selection is common and widespread. But it is not the only mechanism. Many cases of a good fit between organism and environment come about because the organism selects or modifies its own environment. Adaptation is a two-way street; agents of death and destruction impose a regime of selection on organisms in a population, but the organism also chooses the conditions in which it can function and modifies these conditions through its own metabolism and activity. A house fit for a crab or a worm reflects the adaptations and habits of the mollusc that built it; subsequent tenants choose it for its adaptive merits, or improve their fit by employing partners or by adjusting their body to fill the space more effectively. An adaptive whole is a reflection of interactions, feedbacks, and collaborations, a complex entity fashioned not just from natural selection but built through a multitude of simple processes acting together.

Over the very long haul of evolution, secondary shell dwellers have indirectly reinforced the selective regimes that favored the evolution of well-protected housing in the first place. If three out of every four shells are occupied by hermit crabs, as is the case on many tropical shores, the average life span of a shell will be at least three times longer than it would be if all shells belonged to living molluscs. From the perspective of predators that specialize on a diet of shell-dwelling prey, such an increase in the average shell's life span translates into a hugely expanded resource base. This greater resource is an inviting target for still more predators. The predators, in turn, cause the evolution of even more armor among snails, putting still more pressure on poorly defended animals like small crustaceans to seek shelter from predation. An evolutionary feedback of escalation is thus initiated, in which the supply of, and demand for, shells as protective housing increase as shell armor must meet ever more stringent standards set by predators.

When the great seventeenth-century philosopher Baruch Spinoza constructed his rigorously rational conception of the universe and of all the knowledge contained in it, he maintained that everything in this world has a cause, and that nothing could arise without such a cause. This one-way logical link between cause and effect is now being replaced in much of science, and certainly in evolutionary biology, by a two-way link, in which cause and effect influence each other as feedback. Effects and causes flow both ways; they are intertwined to the point where they cannot be separated. There may well be an ultimate trigger for a cascade of events, but the trigger and its consequences are modifed by the interaction between causes and effects.

Ours is a world of interaction at multiple levels of inclusion. A few elements combine in a multitude of ways to create a vast universe of diversity and individuality. This diversity is tamed by selective processes—selective death among cell lines in a developing embryo; selective formation and dissolution of connections in the nervous system, including the learning brain; and natural selection within populations—which distinguish what works from what does not work. Thus diversity, the sum of raw material on which selective processes act, is not merely the consequence of accumulated error, as in the classic view in which biological variation is caused mainly by mutation; it is generated far more dramatically through interaction among elements to create new similar but not identical variations on many themes.

Interaction, then, is the foundation of complexity, and indeed of material existence. Even the most elementary constituents and dimensions of the universe—energy, matter, space, and time—likely represent emergent states that arise through interactions of the kind postulated by string-theory physicists. The basis of all phenomenology resides in elementary interactions. The primary task of science, it seems to me, is to understand interactions and their consequences.

The interactions that give life its unique character are economic. They are not only about how resources are allocated among life-forms, but also about which living things survive and propagate and which do not. Organisms have a major hand in how these interactions proceed, what the criteria for success are, and how history unfolds. The primacy of economics in the affairs of the living world effectively prohibits us from looking at organisms, their parts, and their characteristics in isolation, away from their economic context. It is that context—the sum of interactions in which living things engage—that imparts meaning. Nucleic acids, proteins, genes, cells, organisms, species, and societies are systems and parts of systems, within and among which interactions with economic consequences take place; they are not abstract entities. Nothing about life, evolution, or adaptation makes sense without thinking about how metabolizing bits of life exchange locally scarce resources of matter and energy with only limited access to information and time.

I think back to the curved, tapering tube with a cup coral attached at one end. Its description in the minutest detail is necessary for its interpretation, but description is not sufficient. To make sense of this natural curiosity and to grasp its broader significance, we must place the object in its ecological and evolutionary context, the circumstances and relationships that give it meaning. The simplification that comes with solitary confinement—of names, characteristics, organisms, lineages, or even houses—may make it easier to describe nature, but it precludes true understanding, explanation, and prediction. Nothing in life originates or functions in isolation.

One of the most delightful aspects of the world in which we live is its variation. The economic contexts in which we humans live and in which organisms make their living vary dramatically from place to place and over time. So do the risks that threaten to destroy life and the opportunities that enable life to flourish. The rest of this book is devoted to an exploration of the geography and history of threats, opportunities, and adaptations.

CHAPTER NINE

Dispatches from a Warmer World

There are three indisputable facts about the warming climate to which Earth is being subjected during our lifetime. First, warming is real, rapid, and global, with its greatest effects felt near the poles. Second, it results mainly from a sharp, accelerating, human-caused increase in the levels of greenhouse gases—especially carbon dioxide and methane—in the atmosphere, and is further fueled by runaway feedback processes allowing the planet to retain more of the sun's heat. Third, like many other changes, warming in the short term—on the timescale of human life, from decades to centuries—is on balance harmful, because neither we nor many other life-forms are well adapted to its immediate consequences; it threatens humanity and the rest of living nature with substantial disruption to the status quo. We must either adapt or suffer the consequences.

But there is also a fourth reality: in the long run, warmth coupled with productive ecosystems stimulates adaptive evolution. Ours is not the only time in Earth's history when temperatures worldwide have risen. The fossil record chronicles a fascinating history of climate, including many episodes of warming in times past. Not only does this record reveal how life on our planet was affected by previous warming events, but more importantly it provides us with a long-term perspective on how life responded, a perspective that the brief history of warming over the last few centuries cannot offer. An interesting and largely unexpected finding emerging from studies of the prehuman past is that whereas the short-term consequences of change are nearly always harmful, the long-term effects, which incorporate evolutionary adaptation, are more favorable, enriching life's variety. This conclusion holds as long as other disruptions do not interfere with the temperature-related opportunities that warmth creates. For humanity, the big question is this: can the biosphere in which we live and of which we are a part benefit from this long-term silver lining, or are the convulsions we are visiting on Earth's economy of life—habitat fragmentation and elimination, appropriation of resources to the point of overexploitation, and pollution—so severe that they prevent surviving species from adapting?

If I had not immersed myself in the climate and life of the tropics and delved into the history of climate and life in the geological past, I might never have thought to ask such questions or to hazard such conclusions. Chemistry books told me that the pace of life quickens as temperatures rise from the freezing point of water to a temperature between 95 and 104 degrees Fahrenheit (35 and 40 degrees Celsius). Experiencing the luxuriance of tropical life firsthand, however, I began to grasp the reality that the effects of temperature on life in a forest or on a reef go far beyond the responses of individual creatures. Temperature affects every aspect of the environment, including the ways in which living things interact with one another. In a warmer world, competition intensifies, the range of adaptive options widens, diversity increases, and selection due to enemies strengthens.

Never was the contrast between tropical and high-latitude nature brought home to me more forcefully than during the summer of 1976, when over the course of less than a month I found myself first in the midst of the luxuriant rain forest on Barro Colorado Island in Panama, and then in the equally enticing but much more staid coniferous forest on San Juan Island in Washington State. These back-to-back exposures taught me things that dozens of scientific articles and popular accounts could not convey. Reading about a place is by necessity a sequential act, filtered through someone else's sensibilities. On one page one might read about trees; on another there is a vivid description of the termites or the frogs or the birds. Being in the forest, one experiences everything at once, and one notices things that others overlooked or considered too trivial to mention. Shapes, sounds, smells, and weather come together to offer the prepared mind an emergent conception of the whole. The forest is like a city or the body of a single individual, whose living parts—each with its particular characteristics and relationships—interact with one another and with their inanimate surroundings to create an integrated structure with properties that none of the constituents possesses. The forest is a feast to all the senses, an accumulation of simultaneous, parallel perceptions. As I observe the whole, or stop to examine a leaf or listen to a cicada, comparisons race through my mind, not gradually or serially, but as multidimensional, disordered, yet memorable thoughts. My task as a scientific natural historian is to make sense of all this sensory and mental ferment.

Barro Colorado was once a hill in the midst of unbroken rain forest in central Panama, but when the Chagres River was dammed to form Gatun Lake as part of the Panama Canal in 1914, it became an island. Today it is a forest preserve operated as a research facility by the Smithsonian Tropical Research Institute (STRI), certainly the intellectually most furtive and productive scientific organization ever to have been created in the tropics. Before breakfast in the island's dining hall, one can wake up in the pleasant living quarters to hear the changing of the acoustic guard from crickets and frogs to a chorus of birds and howler monkeys, soon to be joined by a din of cicadas. At STRI's headquarters just a few miles away in Panama City, there are weekly presentations by resident and visiting scientists, as well as the finest library on tropical biology anywhere in the world.

Egbert Leigh was our host and guide on the island. When I first met Bert at Princeton in 1966, he was a beginning assistant professor, a mathematician and theoretical biologist seemingly more interested in thermodynamics and in the equations of population genetics than in the messy world of living things. Ten years later, however, he was applying his mathematical acumen to tropical forests. Not only had he learned the names of most of the common shrubs and trees, but he soaked up everything he could about how these forest architects interacted with the birds and mammals that ate their fruits and dispersed their seeds. Bert was the rare scientist who combines the ability to construct a mathematical theory with a keen sense of natural history. And he did it in the rain forest, one of the most exciting and intellectually impenetrable habitats on Earth.

What, then, do forests reveal about warmth and life? I equate warmth with extravagance and variety in all sensory dimensions. Day and night, the Panamanian forest vibrates to the sounds of a multitude of calls, chirps, buzzes, and songs. Cicadas whine metallically in the canopy by day; frogs pulse at various pitches from the ground and low perches in the evening hours. Birds dueting high above are momentarily drowned out by the shrieks of spider monkeys, sounding too much like children fighting, and the uncouth bellowing and grunting of howler monkeys, all against a constant background of crickets. I am listening to an unrehearsed orchestra of many different instruments playing symphonies and concerti that are at once musically complex and pleasingly transparent. If only the twelve-tone composers of the twentieth century could have produced such lush textures.

Just as pleasing, but more spare and ethereal, are the sounds of the forest of Douglas fir and yellow cedar near the Friday Harbor Laboratories on San Juan Island. The long, low, gravelly croaks of ravens accompany the brilliant soprano singing of wrens, the most plaintive whistling of a white-sided flycatcher, and the owl-like, alto-voiced cooing of a distant fan-tailed pigeon. On rare sunny days, a subtle, delicate buzz of cicadas wafts from the conifers above. This is a soundscape of understated elegance and refinement, more *Eine Kleine Nachtmusik* than a Tchaikovsky symphony.

The architects of the forest—the trees—tell a similar story of contrasts. Hundreds of species vie for a place in the sun in the tropical rain forest, to the point where neighboring trees rarely belong to the same species. The coniferous forest of San Juan Island is no less dramatic. As in Panama, the trees create a moist, wind-still climate near the ground, but in contrast to the enormous diversity at Barro Colorado, there is a sameness to the conifers. Just two tree species—Douglas fir and yellow cedar—dominate the canopy, and many stands contain only one species.

The plants that fascinate me most in the rain forest are the lianes, or woody climbers. Some hang like ropes between trees; others wind so tightly around host trees that their flattened stems incise deep spiral welts on the trunks. Whether they climb with roots or with recurved hooks or by twining around their supports, these plants race toward the canopy to compete with their hosts for sunlight. During my first visit to a rain forest—in the La Selva preserve in Costa Rica—I had noticed that most of the lianes, at least in their understory portions, have heart-shaped leaves, whose broad base makes almost a right angle with the long leaf stalk. Canopy trees, by contrast, tend to have monotonously oblong leaves with a narrow base, with the short leaf stalk inserted in the same plane as the blade. The liane leaves can more easily change orientation to take advantage of the shifting angles of the sun filtering through the canopy. Their broad shape enables the plants to cast maximum shade on the leaves of competitors below.

The tropical profusion of climbing plants makes the almost complete absence of vines in the coniferous forest all the more remarkable. The trunks of yellow cedar and redwood are conspicuously devoid of lichens and mosses, perhaps indicating a chemically hostile bark environment for climbers and epiphytes; but Douglas fir stems are thickly festooned with mosses, lichens, and even small ferns, and would therefore seem ideal for root-climbers; but one searches in vain for vines. Are such plants forbidden in this forest because they cannot grow fast enough to reach the canopy? Does the occasional frost compromise the wide water-conducting vessels in the stem needed to sustain rapid growth?

Comparisons like these might be dismissed as special cases, especially because tropical forests are known to be the most diverse ecosystems on Earth, against which any other systems, including other tropical ones, would look simple. In fact, however, the differences between equatorial and cool-weather forests are replicated in all other ecosystems, notably in the sea. They reveal that, although the rate of production of available food is often highest in midlatitude waters, the range and specialization of adaptive types reach a maximum in the tropics and a minimum in polar oceans. The Guam reef flat, described in chapter 3, and Squirrel Island's rocky shore, mentioned in chapter 1, illustrate these contrasts nicely.

I come away from the geography of adaptation with the indelible impression that the warmth of the equatorial zone brings out an exuberance and flamboyance in its plants and animals that is muted in colder climates. It is as if tropical life operates without the constraints that at low temperatures limit the pace of life and keep extravagant colors, sounds, shapes, and behaviors within a smaller space of possibilities. Warmth is more forgiving of extremes, more favorable to the exploration of new adaptive pathways, and altogether more conducive to the proliferation of species than is unrelenting cold.

At least on the surface, the explanation is simple. Everything an organism does—feed, make proteins and sugars, move, mate, reproduce, invest in offspring, and maintain the body—requires energy. As temperature rises, the chemical reactions underpinning all these activities speed up. Life processes that require copious free energy can take place under warm conditions, when more free energy is available. Many functions become energetically cheaper in the heat. Swimming is less costly in warm water than in cold, because warm water is less viscous, meaning that it offers less resistance to a moving body. Dynamic viscosity of seawater at 50 degrees Fahrenheit (10 degrees Celsius) is 1.27 times greater than at twenty degrees, and 1.48 times higher than at thirty degrees. The formation of mineral skeletons like shells and bones is faster and cheaper under warm conditions because the minerals are less soluble and are therefore more readily laid down in a framework of organic molecules than in the cold. The costs of maintenance increase as the surroundings warm, but these costs are more than offset by savings in most other functions. No thermal limitation prevents tropical plants and animals from leading sedentary lives, but life in the cold is more or less constrained to be slow. At low body temperatures, functions with very high energy demands can be performed only at great expense and with great sacrifice to other necessities, and are achievable only when food stores can be tapped in such structures as bulbs, tubers, or yolks. The higher the temperatures, the more adaptive options are available, and the greater is the realized range of adaptive possibilities.

The high cost of doing business at a low temperature may account for the limited adaptive options available to life in the cold, but the freedom from constraint that warmth offers does not guarantee that the newly available possibilities will be realized. To explain tropical extravagance, we must understand how the permissiveness of warmth turns into evolutionary reality. The lifting of constraint is not enough; it must be accompanied by circumstances that compel exploration of new adaptive domains.

Paradoxically, as the physical restraints on adaptive options relax with increasing warmth, the rules that organisms impose on one another become more stringent. What impresses me most about the tropics, and about life in the hot summers of eastern North America, is that the extravagances of life under those conditions are the evolutionary result of intense competition for necessities that are not only locally scarce, but also under the control of other organisms. Each time individuals encounter each other over some resource like food, mates, or safe places, there is the potential for natural selection, because one party will succeed in acquiring the desired commodity (or retaining it), whereas the other party fails. Encounters among species whose body temperatures mirror those of the surroundings will be more frequent in warm conditions than in the cold, so that individuals in the tropics are tested—that is, subjected to selection—more often.

The signs of greater strife are everywhere. Leaf damage is greater in the tropics; so are nest predation in birds, predation of insects by spiders, shell destruction by fishes and other shell-breakers, seaweed consumption by fishes, and competition for light by plants. As we proceed toward the equator, we find both a higher frequency and greater expression of toxicity in plants and sponges, armor in snails and fish, and risk-minimizing methods of predation that reduce the time needed for a predator to kill and eat its victim. Selection is therefore not only more frequent in the tropics but it leads to greater expression of most traits thanks to the permissive energetics afforded by temperatures between 68 and 95 degrees Fahrenheit (20 and 35 degrees Celsius).

The evolutionary dividends of a warmer world are attainable only if three conditions are met. First, populations must have ready access to a plentiful supply of necessary resources, so that when an imperfect innovation arises, it can linger in the population long enough to be improved by selection. If the population is allowed to grow under a permissive regime of predictable plenty, not every deviant individual is purged from the population, and selection has enough variation to work with. Second, competition for locally scarce resources—the main agency of enemy-related selection—must be intense enough and consistent enough to allow improvements to spread in the population. Third, there must be sufficient evolutionary time—thousands to millions of years—to allow selection to do its work. In order to adapt to greater warmth and to a biologically more demanding world, living things must do more than simply carry on at a higher temperature; they must adapt to, and compete effectively with, other organisms, which likewise are struggling for life under the new thermal regime.

In other words, success in warm surroundings requires adaptations that enable individuals to capitalize on the thermally more permissive environment by becoming better competitors. New, energy-intensive structures and new mechanisms that speed up predation and other risky activities should therefore arise predominantly in warm-adapted lineages that are already under intense selection by members of co-occurring species.

Here, then, is a prediction about history that can be evaluated with evidence from fossils. Shells, which have a prolific fossil record, and many of whose features are readily interpretable in functional terms, are ideally suited for putting this prediction to the test. I needed to identify a simple innovation—a new feature that arose independently in different lineages—that markedly speeds up some important activity in a mollusc's life and that therefore yields a competitive advantage.

I found a nearly ideal group of snails to carry out this test. Among the ways that predators evolved to attack victims that are encased in hard shells, two are particularly widespread in predatory snails. The first method involves drilling a small hole through the victim's shell, and then inserting the feeding organ (the proboscis) through the hole to ingest the soft tissues. Drilling is a time-consuming activity, taking hours or even days to complete. Moon snails tend to subdue their prey while buried in sand, and therefore reduce the risk of detection by their own enemies; but other drillers attack their prey on open surfaces, where potential competitors could easily encounter and interfere with them. The second method used by many predatory snails is to enter the victim's shell through an already existing opening, such as the mouth of a snail shell or the space between the two halves of a bivalve shell. When the victim closes its shell either by shutting the door or by clamping the valves together, access for the predator is difficult, and time-consuming prodding must be used to get inside. Even when access is gained, the proboscis might be severed as the victim tries to deny entry. Injecting a toxin to anaesthetize or disable the victim is one obvious solution to this problem; but wedging the shell, or creating an opening where skeletal elements meet, is another. Many lineages have adopted one or both of these labor-saving methods, which reduce the predator's vulnerability to danger.

It turns out that a small tooth or spine at the edge of the predator's shell lip is an effective tool to wedge open the valves of a clam, to enable the predator to drill or grind a hole at the shell margin, and to prevent the inserted proboscis from being severed. This so-called labral tooth thus cuts the time needed for killing and eating the prey by as much as two-thirds in cases that have been studied. The labral tooth is easily observed on fossil shells, and was therefore well suited as a minor, frequent innovation whose places and times of origin could be inferred by carefully inspecting the fossil record of predatory snails.

Thus began a ten-year search for snails with a labral tooth. I started in my own large research collection, to which I continue to add. Important evidence also came from thousands of species descriptions that I read in eight languages in papers, monographs, and books published as early as 1758; but many of these descriptions were too cursory to be of much use. Had I relied exclusively on the scientific literature for the data I needed, I would have missed dozens of snail lineages that at some point in their history gave rise to species with a labral tooth. I needed to examine more specimens, which often revealed features that others had overlooked.

A particularly memorable instance of this occurred in New Zealand. Before my visit there in 1993, I had become reasonably familiar with its shore fauna, which according to the literature I consulted would contain no species with a labral tooth. I had even queried some colleagues, who assured me that the mudflats and rocky shores I would be visiting were free of such species. I was therefore more than a little astonished when the first shell I picked up on a mudflat on the South Island was a specimen of *Iosepha glandiformis,* an abundant species with a tiny but unmistakable tooth projecting from the middle of its shell lip. Evidently, no one had bothered to remark on the tooth's presence.

This incident reminded me that there is no substitute for carefully examining real specimens. I thus spent countless happy hours examining and measuring thousands of shells belonging to living and fossil species in a dozen European and American museums. Time and again, I uncovered species whose labral tooth had previously been overlooked. Everywhere I went, I compared the tooth-bearing species to related species lacking the tooth, so that I could determine when, where, and in which lineages this predation-enhancing feature evolved. Just as gratifying were the collaborations that blossomed in the course of these investigations. I eventually worked with twelve other scientists on various aspects of the project. Science—even the scientific natural history I practice—is truly a global enterprise, one that, like nature itself, involves as much cooperation as friendly competition in a well-connected community.

Although this kind of work never ends—more and more lineages with a tooth turn up all the time—I currently estimate that some sixty snail lineages evolved a labral tooth. In agreement with the prediction, all lineages with a tooth arose in warm, productive waters, and never in the cold-temperate or polar zones or in the cold deep sea. The tooth would presumably still benefit snails under these frigid conditions—in fact, several tooth-bearing lineages invaded these habitats—but the slow pace of life there might make the advantages too inconsequential for selection to favor the evolution of a time-saving device.

The most interesting result, however, came from the fossils. Times of origin of the tooth were heavily concentrated during four brief time intervals: the Late Cretaceous (80 to 75 million years ago), the Late Oligocene to Early Miocene (25 to 19 million years ago), the Late Miocene (11 to 8 million years ago), and the Early Pliocene (5 to 3 million years ago). Particularly the last three of these intervals correspond to times of conspicuous global warming coupled with widespread productive conditions. The one interval of global warming without the evolution of lineages with a labral tooth is the Late Paleocene to Early Eocene (55 to 52 million years ago). This episode differs from the three later ones in that many marine molluscs remained small, perhaps meaning that seas at that time were relatively unproductive.

Why were there no snails with a labral tooth before about 80 million years ago? It certainly wasn't because seas were cool. One of the warmest episodes in Earth history occurred around 90 million years ago, when tropical seawater temperatures might have been as high as 95 degrees Fahrenheit (35 degrees Celsius)—a full 5 degrees warmer than the maximum average we see in the tropics today—and even polar seas might have been as warm as 68 degrees Fahrenheit (20 degrees Celsius). Despite these truly hot conditions, the main groups of drilling snails had not yet evolved, and the traits of predators and victims alike indicate that predation had not yet escalated to the levels it reached later. Warmth, in other words, may be favorable to energy-demanding innovations to arise, but it is not enough. It is organisms who must take advantage of the enabling conditions to transform potential into reality.

If organisms could maintain high body temperatures even in the cold, they could reap many of the evolutionary benefits enjoyed by tropical species whose body temperatures vary according to their surroundings. This is, of course, exactly what many animals and even some flowering plants can do. Evolutionarily advanced mammals and birds maintain body temperatures of 93 degrees Fahrenheit (34 degrees Celsius) or higher, with some songbirds operating at a regulated temperature as high as 108 degrees Fahrenheit (42 degrees Celsius). Species belonging to groups that diverged early from the mammal and bird branches in the evolutionary tree generally maintain somewhat lower temperatures but are still warm-blooded. Many adult insects—dragonflies, bees, wasps, moths, and large beetles and flies—can raise the temperature of their flight muscles by as much as 30 degrees Celsius above that of the surrounding air, and maintain it during flight. The ability to raise core temperatures to high constant levels has evolved in at least one group of sharks and in various lineages of tuna and their relatives, which are actively swimming hunters. Although their maintenance costs are high, these warm-blooded animals are highly effective competitors. Warm-blooded vertebrates are the competitive dominants on all continents, most islands, and even in most marine habitats, from the shore to the open ocean and the tropics to the poles. At least eight separate groups of flowering plants have evolved the ability to heat their flowers, apparently as an effective means of broadcasting the scent that attracts (and sometimes enforces a night's stay of) pollinating insects.

All these warm-blooded creatures thus evolutionarily tamed, and in some measure have become independent of, the thermal regime of their environment. They have, quite literally, incorporated the opportunities implicit in a high-temperature world into their own bodies, capitalizing on those benefits even where external conditions are thermally hostile. The costs of maintaining these temperatures are, of course, very high, and the use of energy is extremely inefficient compared to that in low-energy life-forms with unregulated body temperatures. As in the case of high-energy humans, whose efficiency of energy use is likewise extremely low, the benefits of warm-bloodedness far outweigh the costs as long as the quantity of available resources is sufficient to sustain such a profligate physiology.

There remains the vexing problem of why so many species live together and evolve in the tropics. It has long been recognized that the number of coexisting species in the tropics exceeds that in comparably large cool-temperate areas by a factor of three or more, and that far more species and lineages originate in regions of warm climate than elsewhere. Curiously, the same geographic pattern applies to human cultures and languages. The greater range of adaptive possibilities at high temperatures implies that there is ample opportunity for organisms to explore avenues of adaptation that are otherwise inaccessible, but it does not explain why adaptive space is filled by so many different kinds of species or cultures instead of by one or just a few entities.

Isolation is the key to species formation as well as to the origins of human cultures. A population of sexually reproducing individuals achieves isolation when a mating barrier separates it from its parent and sister populations. Often this barrier is geographic, a zone of unfavorable habitat across which individuals cannot disperse; but selection can amplify and even create isolating barriers as well. For example, if an ancestral population occupies several habitats, reproduction may take place at different times in each, so that mating is largely between individuals living in the same local environment. This can happen in a parasitic species infecting several host species, or to a population of snails living in vertically separated levels on a rocky seashore. Selection may transform an initially uniform habitat into a patchwork of environments in which the conditions of life differ to the point that descendants become specialized to, and further accentuate, the new regimes. This kind of location selection-driven differentiation has been experimentally demonstrated in laboratory cultures of bacteria. In an initially homogeneous bacterial clone, mutations arise that enable affected cells to form a dense mat, a physically very different environment from the fluid medium in which the original, unmutated cells live. Selection in the two habitats differs, and therefore promotes specialization. Other mutations affect whether certain foodstuffs can be used, and add further dimensions of environmental variation and occupational specialization. Still more subdivision is possible when mutations resisting resident microbial predators or viral parasites arise. Although these bacteria reproduce asexually, lineages under novel regimes of selection create a new environment that is anything but uniform.

In the case of flowering plants, isolation is promoted less by resources than by consumers. Pollinators and seed dispersers faithful to a particular plant species or to a particular type of flower or fruit can quickly help make local mutations into the new norm and promote genetic isolation in their hosts. The same effect applies to specialized herbivorous insects or fungal pathogens. Selection from these sources creates new dimensions of environmental variation and multiplies opportunities for isolation and adaptive divergence. Evidence from both marine and land-based life indicates that selection due to consumers is more intense in the permissive tropics than at the more spartan higher latitudes. The environment of pests, potential helpers, and resources is thus subdivided more finely in the tropics not because of some intrinsic property of the equatorial zone, but because of the selective regimes that organisms are able to impose on each other there.

A similar mechanism seems to be at work among human groups. A common language serves both to unite a group and to distinguish it from others. Groups concentrated near each of several scattered favorable sites may at first speak dialects of the same language, but the dialects quickly diverge to become mutually unintelligible languages. Cultural isolation through language therefore contributes to group differentiation. There is evidence that human groups practicing agriculture or living as hunter-gatherers in forested or mountainous tropical environments are more sedentary than the more nomadic groups on open grasslands at higher latitudes. As a result, cultural and linguistic diversity is markedly higher in the tropics.

Isolation of populations is further promoted by intense sexual selection. As Darwin already knew, the most distinctive attributes of species in which the union between egg and sperm takes place within the body or in an external space controlled by one partner (usually the male) reside in the elaborate reproductive organs and in mating-related displays. There is often intense competition among males for females, as well as discrimination by the female among potential suitors based on the males' weapons, brightness of color, song, or visual antics. Mutations in any aspect of these traits can result in selective mating, and create or amplify mating barriers. There is also abundant evidence for antagonistic coevolution between the male and female organs involved in copulation, where potential injury inflicted by one partner is counteracted by adaptations in the other.

Extremely bizarre reproductive rituals and structures have evolved as a result, especially in insects and in hermaphroditic snails in which mating between partners remains obligatory. Again, mutations involving reproductive structures make isolation of populations easier, and thus contribute to the formation of new species. It is not known whether these isolation enhancements are more common in the tropics than elsewhere.

Still another factor contributing to high tropical diversity is the ability—and often the necessity—for species to maintain themselves as small populations of widely scattered individuals. Following early experimental work by Daniel Janzen in tropical America and Joseph Connell in Australia, Egbert Leigh and his colleagues have emphasized that intense pressure from specialized tropical pests prevents seedlings of trees germinating near their parents from surviving, because the chance that the enemy finds and consumes the seedling is very high near the vicinity of the parent. Low population density of tropical trees can be maintained only because of wide-ranging seed-dispersers and pollinators, which defecate seeds or fertilize flowers of a given species far and wide. Strong selection by friends and enemies in the tropics thus places a premium on distinctive, isolation-promoting reproductive traits. This is yet another indication of how important evolved interdependencies—in this case among plants, pests, and pollinators—is for the evolution of complex ecosystems. If we wiped away the herbivores and the fungal pests, there would be less imperative to maintain a large distance between neighbors of the same species. And if the pollinators and dispersers were eliminated, plants faced with intense pest pressure could no longer maintain viable populations, for they could not afford to be rare.

A warming world, with its poleward expansion of equatorial and temperate climates, enables a large number of species that have already adapted to warm conditions to expand. More intense competition, together with relaxation of the constraints on life's activities, leads over evolutionary time to a greater variety of species. By expanding the opportunities for adaptation, warmth coupled with a dependable and prolific supply of raw materials allows life to flourish and diversify in vigorous profusion.

This long-term enrichment is amply demonstrated in the fossil record. All periods of warming are associated with increases in the local and regional number of species, implying that many new species arise during such times. Tropical lineages, which are usually unable to penetrate higher latitudes because of conservative limits on tolerance of the cold, can break through the thermal barrier and extend into temperate zones during episodes of warming. Cold-adapted species apparently never penetrate the tropics and subtropics no matter what the state of the climate. Warming thus invites evolutionary innovation and diversification.

Inevitably, not every species or lineage can take advantage of these opportunities. Cold-adapted species will be geographically compressed by the expanding warm belt to a narrow band. Their number, however, is relatively small, because there are few cold-adapted species compared to the vast profusion in warm regions. Moreover, the fossil record indicates that few polar species succumbed to extinction during geologically recent warm intervals. The reasons for this are not clear, but one possibility is that, unlike the situation in the tropics where many species are rare, species from polar latitudes tend to be regionally abundant. Barring other disruptions separate from the effects of warming, the loss of temperate and polar species through geographic compression of their comfort zones should be relatively small.

Another potential problem is that many species cannot extend poleward because of unbridgeable barriers unrelated to temperature. Terrestrial species in southern Australia or South Africa would be forced into the sea if their thermal environment is eliminated by global warming. In the human-dominated world of fragmented habitats, many species have become restricted to islandlike patches from which escape is all but impossible without assistance. Rising sea levels, changing patterns of precipitation, and other changes accompanying warming may further erect or broaden barriers to the poleward spread of land-dwelling and freshwater species.

Why should the short-term effects of warming often be harmful when the long-term consequences are on the whole beneficial? The answer comes down to adaptation. If an organism, population, or ecosystem is well suited to its situation, almost any change will make it less so. This is why most genetic mutations are harmful, why many ecosystems are disrupted when species are introduced to them, and why many workers suffer when new labor-saving methods are put in place. If, however, the system is able to adapt, or to accommodate to the changes, the short-term deficits can give way to long-term benefits. For example, mutations are an important source of variation, which is a necessary raw ingredient on which selective processes can work to fashion a better-adapted whole. The spread of species into communities they did not originally occupy can add resources and regulation to those communities, and becomes essential for reconstituting ecosystems that have been devastated by catastrophe. Change is universal. Those organisms, lineages, ecosystems, and societies that can accommodate it flourish while those that cannot will either wither away or become marginalized. Evolution, in other words, can transform adversity into opportunity.

Whether warming is a curse or a blessing thus depends entirely on whether living systems subjected to it can adapt or move. If they can, warming presents an opportunity, especially if the surroundings are healthy and productive. If they cannot, warming becomes a hardship, an insuperable challenge. Humanity should do what it can to limit the rate at which the world is heating up, but above all we must adapt to a warming world. If we want to maintain some semblance of wild nature in the face of warming and habitat fragmentation, we must preserve—or, better yet, enhance—opportunities for species to adapt. We must give them wiggle room, not box them in. We must allow evolution and adaptation to do their work.

CHAPTER TEN

The Search for Sources and Sinks

Africa holds a very special place in the history of our species. The consensus of scientific opinion has it that our hominid lineage, as well as the modern human species *Homo sapiens,* originated on that great continent. Although some anthropologists favor Asia as the cradle from which people spread throughout the rest of the world, the bulk of molecular and fossil evidence pinpoints Africa as our birthplace. Regardless of whether Africa or Asia will prove to be the land mass where we evolved our unusual characteristics and the technology that fueled our world dominance, it is important to discover whether these continents and their ecosystems possess some quality that made them more likely sources for a species like ours than, say, the Americas or Australia. How do they differ from other land masses, and do such differences matter? How, in other words, does place affect adaptive evolution?

North America and Australia can be ruled out as sites of human origin because of the absence of suitable primate ancestors. The island continent of South America, however, harbored a vigorous and diverse group of New World monkeys, which arrived there from Africa some thirty to thirty-five million years ago. The potential existed on that continent to evolve the functional equivalent of the great apes of Africa and Asia and, ultimately, to evolve a humanlike primate. Even Madagascar, with its diverse assemblage of lemurs, would not have been out of the running on account of suitable progenitors, because a giant ground-dwelling lemur with apelike characteristics thrived on this island before the first human settlers arrived there about a thousand years ago. Quite apart from the presence of primates, would the selective regimes in settings like Madagascar, Australia, or North and South America have been conducive to the origin of the intellectual, locomotor, dextrous, social, and technological attributes that make *Homo sapiens* the overwhelming ecological dictator of the world?

Scientists are ill-equipped to answer "what-if" questions, but we can rephrase them in ways that allow for more fruitful advances of inquiry. The origin of the hominid lineage and the origin of modern humankind are unique events that cannot be replicated, so that an explanation for those particular historical transformations would remain a "just-so" story, a narrative for which many separate lines of evidence could be adduced but for which no general principles could be extracted. If, however, our evolutionary twig is taken to represent a broader class of lineages—in this case, species that achieve economic supremacy in the larger systems they occupy and build—then it becomes part of a larger sample, other members of which can be investigated and compared. If the particulars of human origin have something in common with the origins of other lineages that have reached positions of power either today or in the past, a systematic, statistical approach guided by expectations from evolutionary theory can reveal common circumstances.

The broader question raised by such inquiries is this: are there regions, habitats, or ways of life that are more conducive than others to the evolution of such influential traits as a high level of activity, large body size, complex social organization, virulence (or infectiousness) of pathogens, resistance to communicable diseases, symbiosis between cooperating species, and weediness? A very similar question has occupied economists and historians from Adam Smith onward: which circumstances enable some societies to industrialize and to become wealthy while others remain at, or in a few cases regress to, subsistence levels? What, in other words, is the geography and evolutionary distribution of opportunity?

I was naturally drawn to these questions. Besides my attractions to all things natural historical and scientific, I shared with my brother Arie a fascination with geography. With his considerable drawing talent, Arie constructed detailed Braille maps for me by tracing rivers, coastlines, and borders on a thick sheet of Braille paper laid on a piece of metal window-screening that my father had nailed to a sheet of plywood. By pressing firmly with a Braille stylus over tracings he had done by pencil, my brother created a tactile map on the reverse side of the paper, which he then proceeded to fill with Braille letters and numbers denoting cities, mountains, and any other geographic details that the atlases he had at his disposal revealed. The names of faraway places were to us the stuff of daydreams. They appealed to me as much as did the Latin names of the plants and animals that were beginning to absorb me more and more. The idea of combining my budding biological interests with geography fired my imagination. Making it come true would have to wait, but meanwhile Arie and I read all we could about distant places and their natural environments and inhabitants.

Simply associating geographic locations with the names of species would never be enough for me. I longed to observe life in unfamiliar places, to glimpse the shores and reefs where the makers of the beautiful shells I was being given from Australia, the Philippines, Zanzibar, and Florida lived. I came to understand early on that the geography of life is the geography of adaptation.

Curiosity grew to scientific engagement when, as a graduate

student, I read papers by John C. Briggs, a fish specialist with a strong interest in the relationship between geography and evolution. Like others before him, including Charles Darwin, Briggs held that the competitively most vigorous species evolved in regions of large size where the number of coexisting species was high. As a marine biologist, Briggs was particularly impressed by the richness of life in the archipelagoes and seas of the so-called Indo-Australian region, comprising the Philippines, the mainland coast of southeast Asia, and the shores of Indonesia, New Guinea, and northern Australia. This diversity hotspot supports tens of thousands of marine species, far more than any other marine region on the planet. A biologist outfitted with only a snorkel can find perhaps a hundred species of reef-building coral on a single day in Palau, a small archipelago of stunningly beautiful islands in the tropical western Pacific. By contrast, the entire Caribbean Sea contains no more than sixty species of coral. Briggs suggested that the Indo-Australian region was the site of origin of many species, which subsequently spread throughout the Pacific and Indian Oceans and often to other parts of the marine tropics. This geographic vigor, Briggs argued, was attributable to the intensely competitive conditions and high local diversity prevailing in the innumerable reefs, mangrove forests, sandflats, rocky-shore habitats, and deep-water hardgrounds of this tectonically active region.

Everywhere we look, we find surprising and often baffling differences in the ecosystems of places that, on the basis of climate, appear similar. Rain forests of Southeast Asia and tropical America are richer in species than those in tropical Africa. Trees emerging above the forest canopy are more common in Southeast Asia than in rain forests on other locations. Islands differ from continents, and the ancient river systems of North America—the Mississippi and its tributaries, and the other rivers draining into the Gulf of Mexico—support a far more diverse fauna of clams, snails, and crayfish than their counterparts in Europe or even East Asia. These contrasts seem to correspond to geographic differences in the duration or intensity of selection to which resident populations have been exposed.

These ideas jumped off the page in Panama. Nowhere else on Earth do two oceans—the Pacific and the Caribbean part of the Atlantic—come so close while still being separated by a continuous land barrier, which in that part of Central America is about fifty kilometers wide. On several occasions I was able to visit shores of both oceans on the same day, an experience that hammered home the dramatic contrast between the two coasts. The Caribbean Sea near Panama is warm and nearly tideless, and its clear, relatively unproductive waters support lush coral reefs, mangroves, and sea-grass meadows. Panama Bay, on the Pacific side, is much cooler, especially when deep nutrient-rich waters come to the sea surface during the dry season from December to April. Its waters are murky and full of plankton, and there is a huge tidal range of four meters or more. Enormous sandflats, mangroves, and boulder shores teem with life, but reefs are small and local, and sea-grass meadows are perplexingly absent. The Pacific coast of Panama harbors some of the best shell beaches in the world.

Although my interest in this Pacific-Atlantic contrast was aroused by a general curiosity about the geography of form in seashells, it took on a more practical dimension as the possibility of building a sea-level canal across the Isthmus of Panama became a subject of intense scientific debate. The current freshwater canal, connecting the two oceans through a series of locks, was unable to accommodate supertankers and very large container ships. A wider sea-level canal was therefore economically attractive and technically feasible. Environmental concerns played no role whatsoever when first the French, and later the Americans, planned and executed the construction of the freshwater canal. But now, in the 1970s, biologists wondered what would happen to marine plants and animals on the Caribbean and Pacific sides of Panama if a sea-level, saltwater channel were to replace the river-fed freshwater canal, which has acted as a highly effective barrier to the interoceanic dispersal of marine species.

Briggs had predicted that, because Caribbean fish communities were more diverse than their eastern Pacific counterparts, the number of Caribbean species that could spread to the Pacific through a sea-level Panama Canal would exceed the number migrating in the opposite direction. Other scientists, however, worried about the arrival in the Caribbean of Pacific species that had no ecological counterparts in Atlantic waters. One species of special concern was the crown-of-thorns starfish. This species—actually several species, according to recent molecular findings—is aptly named for its sharp venomous spines. It is a voracious predator of reef corals, with an enormous geographic range from the Red Sea and East Africa across the Indian and Pacific Oceans to the west coast of tropical America. Nothing like this sea star exists, or has ever existed, on the reefs of the West Indies and Brazil. Should this sea star succeed in colonizing the Atlantic, the tourist economies of many Caribbean nations could be imperiled as outbreaks of the sea star devastate the reefs.

And then there is the ocean-going, aggressive, venomous sea snake, another species with a vast range in the Indian and Pacific Oceans. Experiments by Ira Rubinoff—soon to become the director of the Smithsonian Tropical Research Institute—and an Israeli postdoctoral student had shown that, whereas large Pacific fish avoid attacking and swallowing these snakes as prey, naive Caribbean fish with no prior experience readily ingested them. The snakes would remain alive long enough after being swallowed to bite their attackers from within the body. Atlantic fish might soon learn to associate snakes with mortal danger, but meanwhile the arrival of such novel predators by way of a widened and deepened Panama Canal could pose a real threat.

My perspective on the sea-level canal came from snails. Taken together, the shallow-water assemblages of species from the eastern

Pacific contained snails with larger, more heavily armored shells than their Caribbean counterparts. Together with a large contingent of coral-eating predators that had no Caribbean equivalents, eastern Pacific species might on average hold a competitive advantage over native species in the western Atlantic. On these grounds, I suggested that species would predominantly move from the Pacific to the Caribbean through a sea-level canal across Central America.

A precedent for this kind of invasion was already well known to marine biologists. Upon completion of the Suez Canal between the Red Sea and Mediterranean Sea by French engineers in 1869, hundreds of species from the Red Sea began to invade the eastern Mediterranean, whereas few if any species extended their range in the opposite direction. Among the new arrivals are several fish—for example, the red squirrelfish, a filefish, and a cornetfish—that have become spectacularly abundant, as well as the sharp-spined sea urchin *Diadema setosum*. These invaders have begun to enrich the fauna of the eastern Mediterranean, which before the opening of the Suez Canal had fewer than half the species known from the northern Red Sea. Moreover, the native Mediterranean molluscs had much less heavily armored shells than did invaders from the Red Sea.

Fortunately, none of the predictions about the biological consequences of building a sea-level canal across Central America has had to be tested. I say fortunately because uncontrolled experiments of mass introduction of foreign species are ethically irresponsible, given the damage some of the species could inflict. A sea-level canal is, at least for the time being, not in the offing, although its construction is well within current technical reach. Instead, a widening of the existing freshwater canal is underway in Panama, together with the construction of an additional set of locks on the Pacific side. Transfers of species from one ocean to another through natural dispersal will therefore remain very limited. Nevertheless, the ecological and evolutionary questions raised by projects like the Panama and Suez

Canals will not go away. It remains important to understand which species are most likely to disperse from one geographic region to another and how such migrations affect recipient ecosystems.

Human engineering is only the latest agency to alter world geography and to bring about dramatic changes in the distribution of plant and animal species. Climatic changes, as well as geological forces causing tectonic plates to splinter and collide, have opened and closed ocean gateways, modified the pattern of current flow to the oceans, and erected and severed land bridges. Careful studies of fossils, ancient climates, and the restless crust and ocean floor, together with inferences about the patterns of descent and branching in the evolutionary tree of life, have enabled paleontologists to piece together the eventful historical geography of life.

The picture that emerges is one of frequent episodes of species expansion from one geographic region to another. The pattern of spread is almost always strongly one-way, with one region acting as the source, or donor, and the other as the recipient, or sink. Lineages expand from continents to islands, from shallow to deep waters, from surface habitats to caves, and from the tropics to higher latitudes. They cross oceans, move between continents, and spread from one ocean to another. Why are these patterns of invasion so lopsided, so strikingly asymmetrical? What do donor regions have that recipient regions or habitats do not?

It was this question that led me out of the tropics into the cold northern oceans. As I mentioned in chapter 1, the opening of the Bering Strait and a relatively warm Arctic Ocean enabled hundreds of species to expand from the North Pacific to the North Atlantic in the so-called trans-Arctic interchange. In this highly lopsided movement of species, the North Pacific served as the predominant source region and the Arctic and North Atlantic Oceans as the primary recipient regions. Compared to the North Atlantic, the northern reaches of the Pacific Ocean, including the Bering Sea, are more productive, three times richer in species, and home to larger, more powerful predators—crabs, sea stars, and carnivorous snails. Uniquely among cool oceans, the North Pacific supported a large herbivorous marine mammal, Steller's sea cow, which became extinct during the eighteenth century. In short, the seaweeds and animals of the North Pacific had evolved under a regime of intense competition and predation, a setting in which the standards of performance and survival and propagation were very high.

Now that global warming is rapidly diminishing the extent and thickness of Arctic sea ice, conditions in the Arctic Ocean are reverting to those prevailing before the mid-Pleistocene transition. Current estimates indicate that the Arctic Ocean will be seasonally or perhaps even perennially ice-free by the year 2050. From the historical record of life in the warm Arctic, we can expect a resumption of widespread trans-Arctic transport of temperate species from the Pacific to the Atlantic. Because few marine invasions lead to the extinction of species in recipient regions, the projected trans-Arctic expansion of Pacific species should enrich North Atlantic ecosystems and raise the competitive bar in them. Though certainly not an experiment I would have wanted to perform, the events forecast for the northern oceans of the future will be interesting to observe as a natural experiment initiated by human activity.

Species have also been on the move within both the North Pacific and North Atlantic Oceans. Working with my Japanese paleontologist colleague Kazutaka Amano and with several others, I came to understand that a significant fraction of the North Pacific fauna (and probably also flora) originated in the warm-temperate waters of California between thirty and twenty-five million years ago. During an exceptionally warm interval from seventeen to sixteen million years ago, many elements of this eastern Pacific fauna extended northward and westward across the Pacific to northeastern Asia. Later, during the lesser warm spells ten million and three-and-a-half million years ago, the direction of expansion seems to have reversed, with Asian species now extending back toward North America. In the North Atlantic, fossil evidence as well as genetic studies show that Europe has served as a consistent source for species and lineages that today live on both sides of the Atlantic. The reason for these patterns of spread remains unclear.

Perhaps the most celebrated case of species exchange among continents is the so-called Great American Interchange. Movements of species between North and South America were nearly impossible when the two continents were separated by deep seaways, but as these passages narrowed, raccoons and perhaps some rodents were able to colonize South America from the north. A trickle became a flood of invaders when the Central American isthmus became continuous, probably as the result of falling sea levels associated with the expansion of ice in the northern hemisphere about three million years ago. North American carnivores—weasels, dogs, cats, and bears—overwhelmed native South American predatory marsupials and other mammals that had evolved under less stringent island conditions. Herbivores including elephants, camels, and peccaries also streamed into South America. Many of these invading lineages diversified in South America. In an odd twist, North American camels became extinct on the continent where they first arose, but survive today in South America as well as in Eurasia and North Africa, areas to which camels had spread by about two-and-a-half million years ago. Relatively few South American mammals moved to North America, and those that did generally remained in the warmer parts of that continent and did not diversify. Thus sloths, armadillos, anteaters, and New World monkeys came north, but they did not extend beyond the southwestern United States and Florida. Among mammals invading from South America, only the porcupine expanded to high northern latitudes. Forest-dwelling plants and animals moved north from South America, whose tropical area greatly exceeded that of North America.

Among the several potential causes for this southward march of mammals, the most important seems to be the competitive superiority of the northern invaders. As a much larger continent, North America produced not only larger top predators and herbivores among its mammals, but also animals with higher metabolic requirements and thus a higher demand for food. Before the Bering Strait opened, North America was connected with Asia, and thus would have provided an even greater extent of continuous land for large, actively roaming carnivores and herbivores. With the establishment of a continuous Central American corridor, North American invaders changed the island continent of South America into an extension of the North American land mass, and through conquest raised the competitive bar there.

The closing of the Panama seaway may have been a boon to land mammals taking part in the Great American Interchange, but it was disastrous for many marine species. One of the first scientists to grasp the full extent of this disaster was Wendell Woodring, a paleontologist at the U. S. Geological Survey who, upon retirement, set about publishing six large monographs on the fossil molluscs of Panama. By the time I met Woodring at the Smithsonian in the mid-1970s, he had already had fifty years of publications behind him. Though elderly, he was still physically and mentally vigorous, with a distinctly refined air about him. On a visit to Panama in 1976, where he participated in a meeting organized by Egbert Leigh to talk about the geological and biological history of Central America, Woodring climbed the more than two hundred steps from the boat landing on Barro Colorado to the clearing where the research facility and housing were located.

Woodring and his contemporary Axel Olsson, who also attended the Panama meeting and who had made much of his reputation in Florida, Ecuador, and Peru, were part of a long tradition in paleontology, in which most advances were made by scientists working alone. Jeremy Jackson, whom I first met as a graduate student at Yale in 1968, was convinced that this lone-wolf tradition was ill-suited for tackling really big problems, such as understanding the highly complex geological and biological history of Central America, an area where forces in the Earth's crust changed a chain of volcanic islands into a continuous strip of basins and mountain ranges over the course of some fifteen million years. A tall man with a sonorous voice, a deep-throated, almost ghoulish laugh, and a powerful intellect, Jeremy assembled an international team of geologists, paleontologists, and molecular biologists to form the Panama Paleontology Project, headquartered at STRI. Although I was never formally a part of this effort, several of my students and I have greatly benefited from the collections and findings that the group made. Moreover, I had made several visits to the Gatun Formation, an incredibly rich fossil deposit on the Atlantic side of the divide in Panama, where Tony Coates, the chief geologist of the project, served as a highly knowledgeable guide. The Gatun Formation is one of those glorious deposits where fossil shells are so plentiful and so exquisitely preserved that it was like being on a particularly good shell beach. It is the perfect setting for a lazy paleontologist like me, who is content to collect fossils without the aid of picks, shovels, and hammers.

As Woodring already recognized in 1966, hundreds of marine molluscan lineages with representatives in both the Caribbean and Pacific sectors of the New World tropics when the seaway was still open suffered extinction, but the Caribbean was much more affected than the Pacific. In the latest tally, conducted by the Portuguese paleontologist Bernard Landau and me, sixty-eight snail lineages became restricted to one side of tropical America or the other. Of these, sixty-four became what Woodring called Paciphiles—lineages that became extinct in the Caribbean and survive in the eastern Pacific—and only four died out in the Pacific and still live in the Caribbean. The Pacific coast therefore became a major refuge, a safe haven where many of the species that lived twelve to eight million years ago during Gatun time had left living descendants.

One group of species that was particularly hard hit by the emergence of the Panama land bridge comprised molluscs with a larval stage that fed and grew while living in the plankton. Without a plentiful supply of planktonic food, populations with such larvae are unsustainable. One of the most important findings to come out of the Panama Paleontology Project is the discovery that, whereas plankton became more abundant in the eastern Pacific as the isthmus was forming, largely because deep waters rich in nutrients welled up to the ocean surface, large parts of the Caribbean Sea were becoming increasingly starved of planktonic food. On the shallow Caribbean seafloor, animals filtering food particles out of the water were being replaced by corals and other reef-associated animals, many of which possess photosynthesizing microbes in their light-exposed tissues. Only the Caribbean coast of Colombia and Venezuela in northern South America remained planktonically productive thanks to upwelling that has been in place for at least nineteen million years. Yet even on this coast of plenty, many lineages—at least 11 percent of snail lineages that were living there three-and-a-half million years ago—became extinct. It was therefore not just the local food supply that placed populations in jeopardy, but also the food supply in those critical sites where a disproportionate number of larvae were being produced. In other words, the fate of entire species depended on the nutritional health of source regions, where the bulk of individuals in the next generation was being born.

Source populations and regions thus turn out to be crucial to the survival of species and for defining the adaptive characteristics of species. Most members of a species are born in source populations, which occupy the most productive habitats and regions within the geographic range of the species. The selective regimes in these cradles thus dictate the adaptive characteristics of the species. Sink populations, those which cannot sustain themselves in the absence of recruits from elsewhere, contribute little. If they are cut off from their sources, these populations are doomed to eventual extinction unless something happens to transform the sinks into sources. This can occur if the food supply or the size of the habitat in the sink region increase, or if per-capita demand is diminished, so that a given small habitat can support a larger population. For example, isolated island populations of large mammals are not viable, but they often evolve toward smaller adult body sizes. This must have happened on the Indonesian island of Flores, where immigrant stegodont elephants evolved into a dwarf species, and an early hominid evolved into the miniature *Homo floresiensis,* both now extinct. On the Mediterranean islands of Cyprus, Crete, and Sardinia, among others, the body masses of dwarf island elephants, deer, and hippos dropped by half or more after their normal-sized ancestors arrived in the Pleistocene. Although all these island mammals have become extinct, likely as the result of human predation, their dwarfing and the associated lower per-capita demand for food enabled populations of these species to persist for thousands of years.

Bernard Landau and I interpret the extinction pattern in marine tropical America in this source-sink framework. Before the Central American isthmus became an impassible barrier, currents flowed eastward from the Pacific into the southern Caribbean, bringing plankton-dwelling larvae from the productive Pacific with them. As the barrier rose, the increasingly plankton-starved Caribbean was progressively cut off from the increasingly plankton-rich eastern Pacific. For species dependent on a plentiful planktonic food supply for the larvae, populations in the Caribbean became sinks and eventually declined to extinction. Some species whose larvae can remain in the plankton for many months may be able to subsist on a low-density diet of plankton; these species were able to hold on longer in the Caribbean.

As our speculations about the Caribbean make clear, the sites that act as sources and that produce a surplus of individuals can change location. Conditions in the new source region are unlikely to be the same as those in the old, meaning that the selective regime—enemies, allies, and weather—and therefore the adaptations of the species could also change when one source region is replaced by another. The characteristics of individuals in a species are determined by majority rule—that is, by a selective regime to which the majority of individuals of a species is exposed.

My long-time friend and colleague Gregory Dietl, at the Paleontological Research Institute in Ithaca, New York, and I applied this concept of majority rule to explain the widespread evolutionary phenomenon that Niles Eldredge and Stephen Jay Gould called "punctuated equilibria." Eldredge, a paleontologist at the American Museum of Natural History in New York, had observed in 1971 that trilobite species living in North America during Devonian time, 410 to 365 million years ago, remained unchanged in form for millions of years, and then suddenly—over a period of perhaps as little as a few thousand years—evolved into new species. Earlier work by the Dutch oil-company geologist Henry MacGillavry in Indonesia revealed a similar pattern in large, seafloor-dwelling, one-celled Foraminifera. Constancy thus alternated with periods of rapid change in what appeared to be a punctuational, stop-and-go pattern of evolution. Dietl and I proposed that a geographic relocation of source populations coupled with a corresponding change in selection and therefore in adaptation of the majority, is sufficient to account for punctuated evolution.

If we wish to understand the origins and adaptive histories of influential organisms, we must look to the productive environments and regions that support source populations of many species. It is in these situations where adaptive specialization is least constrained and where the potential for evolving large body size, high metabolic rates, complex social organization, risk-minimizing means of predation, weedy opportunists with high fecundities, and other high-energy means of making a living is highest. Relentless selection for high performance exists always and everywhere, but it can proceed further in large habitats where plentiful nutrients are readily accessible. Nowhere else can low-density populations of high-energy individuals, such as voracious top predators and rapidly metabolizing herbivores, maintain viable populations. High-performance dominants impose intense selection on all other species, and are thus indirectly responsible for the evolution of parasites, virulent pathogens, and mutually beneficial symbiotic associations between microbes and their hosts. Warm-blooded vertebrates and sophisticated insect societies originated on the continents, not on islands; so did venomous snakes, large herbivorous land snails, herbivore-resistant grasses, and tall forest trees. The largest, most voracious marine crabs, snails, mantis shrimps, sea stars, and reef fish are found in shallow-water habitats along the shores of continents and large high islands, and typically have huge geographic ranges. In the more limited habitats around small isolated islands, individuals of large-bodied species in the various feeding categories—herbivores, carnivores, suspension-feeders, and deposit-feeders—are simply absent.

The adaptations of competitive dominants on isolated islands revolve around low growth rates, long life spans, limited mobility, passive defenselike armor, and low fecundity. For example, the top herbivores on the Galapagos Islands off South America and Aldabra in the Indian Ocean are large, one-thousand-kilogram, slow-growing, long-lived, sluggish tortoises with a heavily armored carapace. On most other islands, the role of top herbivore is (or was) filled by large flightless birds, such as the moa in New Zealand and the moa-nalo (related to geese and ducks) on the Hawaiian Islands, both now extinct. Flightlessness, in fact, characterizes dozens of island birds—rails, geese, parrots, cormorants, ducks, and even the

New Zealand wren—as well as insects. The slow pace of life indicated by these adaptations suffices on islands undisturbed by humans, but on the continents and in large expanses of productive shallow sea it works only for species in economically subordinate positions.

All islands of the world have been biologically altered to one degree or another, both through the extinction of native species and the introduction by humans of species from the continents. The success of introduced species on islands poses a puzzle: why don't species with characteristics that evolve on continents evolve on islands? The answer is to be found on the island as well as in the unsuitable habitats separating the island from potential source areas on the continents. Insofar as islands are economies unto themselves, isolated from systems elsewhere, they are small units where high-energy lifestyles cannot be maintained. If a high-powered immigrant is to thrive on the island, it will have to reduce per-capita demand, as happens in the dwarfing of mammals, discussed earlier. The big problem, however, is that most high-energy species cannot colonize islands. In order to cross the water barrier, most would-be colonists must either float by themselves or be carried on some vessel like a log, bird, or bat. During the passage, colonists must remain largely inactive, for they cannot feed or make food. This imposed inactivity precludes the high-demand adaptations that characterize so many vigorous continental species. Only flying animals are more or less exempt from this restriction, but weight limitations associated with long-distance flight constrain these animals to be small. Far-off islands are thus unlikely to be reached by large birds, mammals, or trees. Those low-energy species that do reach islands can, and often do, increase power as they adapt to the new conditions, but they never attain the productivity and per-capita demand of species that evolved in the larger economies of continents. The ten-foot-long Komodo dragon, a fierce predatory lizard on several islands in southeastern Indonesia, is larger than its presumed immigrant ancestor, but its cold-blooded metabolism ensures that this local top predator has only about one-fifth the food requirements of a continental predatory mammal of similar weight.

The evolutionary perspective on sources, sinks, continents, and islands encourages us to think about our ecological future in a new way. That future consists of still more exploitation of the resources that remain—forests, fisheries, soils, the ocean plankton, and many individual species—and of more fragmentation of ecosystems that were once much larger. Exploitation has traditionally targeted the most productive environments, the very settings where most source populations are concentrated. Conservation efforts, by contrast, have targeted ecologically marginal habitats, which cannot support economically significant agriculture, wood production, or fisheries. Evidence and theory, however, strongly indicate that it is source regions and source populations—large, productive environments—that deserve our highest priority for protection. This means saving what is left of lowland forests, grasslands, coral reefs, mangrove swamps, salt marshes, continental shelves, upwelling regions, sea-grass meadows, kelp forests, large productive rivers and lakes, estuaries, and the open ocean's plankton. I am, of course, not suggesting that the mountains, deserts, polar environments, deep sea, and isolated islands that have for one reason or another been spared the ravages of human exploitation and disruption should now be targeted for destruction. It is important both morally and aesthetically to protect these environments as well, but we can avoid the most far-reaching ecological meltdown only if serious efforts are made to halt or reverse development in the productive ecosystems on which our own long-term well-being depends.

The fragmentation of "natural" ecosystems is transforming Earth's biosphere into a patchwork of islandlike habitats, in which all but a few tramp species—weedy species and other animals and plants like sheet grass, salt-tolerant tamarisk, rats, starlings, and crows that are closely tied to humans or to human-disturbed environments—will experience a dramatic shift from a selective regime characteristic of continents and large ocean ecosystems to an insular setting with its lowered competitive standards. This trend toward the splintering of ecosystems will exacerbate the effects of targeting productive source regions for exploitation. If we are to slow or reverse this trend toward insularization, we must heed the historical record and shift our emphasis from protecting single species, sometimes under artificial conditions like zoos, to large tracts of unexploited habitat.

In this light, human origins in Africa make sense. Africa and Asia were already home to the world's largest warm-blooded predators (lions and tigers), the fastest mammals (cheetahs), the largest tropical herbivores (elephants), and some of the most aggressive snakes. On these very large land masses, which had been exchanging evolutionary lineages for at least twenty million years so that their ecosystems were effectively united, performance standards had been driven very high indeed. Competitive dominants in such a situation must have exceptional capacities to be successful. The threat of predation by social hunters like lions and hyenas would have been high, especially for vulnerable young; and competition for food would have been fierce. Our unique qualities—high levels of cooperation, intelligence, mobility, protection of offspring, and the use of tools including long-distance weapons and fire—were useful extensions of traits that had their prototypes in our earlier primate ancestors in Africa.

Why, then, did a humanlike species not evolve in North America, which was also a large continent, made even larger when it was connected to Asia during most of the last fifty-five million years of mammalian evolution? The easy answer, as I suggested at the beginning of the chapter, is that there was no suitable primate ancestor in North America, precluding one potential avenue for a highly intelligent, social animal to evolve. But other avenues might have existed, and the complement of native North American mammals—dogs, large cats, and bears—was not so different from that in Africa, and would have exerted selection favoring group defense, offensive weaponry, and other humanlike traits in a suitable victim species. In time, North America might well have spawned a species like ours. The fact that it did not, whereas Africa did, illustrates the role of chance in evolution, a topic to which I shall return in chapter 12. Over the long term, a creature with humanlike traits is almost sure to emerge, but when and where this will happen is in large part a matter of chance.

These biological ideas may seem worlds away from modern human affairs, but they are not. In *The Wealth of Nations,* Adam Smith anticipated many of these ideas when he wrote about the contrasts between rural and urban life. In his conception of rural settings, each worker must accomplish many tasks, and be a jack-of-all-trades. There is little direct competition among workers, and therefore no strong incentive to specialize in specific types of work. In the larger economies of the towns, individual workers become tradesmen, clergymen, weavers, butchers, moneylenders, innkeepers, and the like. Life proceeds at a faster pace and gives rise to occupations that simply did not exist in the countryside. It is also in the towns where prototypes of the great corporations arose. Early versions in the late medieval period in Italy and the early seventeenth century in the Netherlands were cooperative ventures focused on overseas trade, which not only enriched the towns but also brought new commodities into the marketplace of town and country alike. The economies of the towns were ideal environments for entrepreneurs to invent new financial structures—banks, stock markets, and the bond market—and for inventors to design labor-saving machines. Great advances in mathematics and science were centered in towns from Paris and Pisa to Florence, London, and Amsterdam. Towns were therefore engines of wealth creation, much as the ecosystems of large continents and oceans are.

As in nature, none of this can happen without an initially productive economic base. The civilizations of Peru, Egypt, and Mesopotamia arose near coasts with prolific marine resources. Fertile soils in river valleys and on great river deltas produced an agricultural surplus, which was critical to the emergence of urban centers and trade in Egypt, the Fertile Crescent, the Indus Valley, and coastal China. The same combination of highly productive agriculture and abundant fisheries enabled the Netherlands to emerge as a disproportionately powerful trading nation in the seventeenth century. The economies of cities, nation-states, and international corporations are thus like the productive ecosystems of great land masses and shallow seas. It is the activities of individuals and organized groups that transformed small, isolated, slow-paced economies into large vigorous ones; and it is the added value brought to an economy by trade and innovation that builds on natural plenty to produce vibrant wealth.

But these centers of wealth and power are also bastions of what can be characterized as adaptive and economic complacency. The threat of invasion and revolution appears, at least for the time being, to be contained. The problem with such self-satisfaction, or safety in effective adaptation, is that the symbols of outward success—wealth, power, and the status of economies as sources—are subject to disruption and usurpation. Power is not permanent; it flits notoriously from place to place, from one party to another. In the next chapter, I look at how power both promotes and interferes with evolutionary opportunity.

CHAPTER ELEVEN

Invaders, Incumbents, and a Changing of the Guard

It is often said that familiarity breeds contempt, but in the realm of natural phenomena it seems more often to breed apathy. Precisely because something is an everyday experience, we think we understand it, especially if we give it a name. There is no puzzle to be solved, no mystery to be unveiled. Yet it is often the ordinary things that are least understood. The job of a scientist is to recognize quandaries overlooked by most others, and to ask questions in such a way that explanations and answers can be discovered.

One overly familiar bit of natural history that attracted my attention is the rarity of insects in the sea. With at least a million species of insect in every imaginable habitat on land, and an additional forty-five thousand in freshwater lakes and streams, there are only about fourteen hundred that live on marine rocky shores, on mudflats, in salt marshes, and—in the case of five species of water strider—on the surface of the open ocean. These invaders to the sea represent perhaps 120 separate evolutionary lineages, about two-thirds of them coming directly from the land and the remainder entering the sea from freshwater. What is it about insects that prevents the overwhelming majority of their lineages from penetrating the marine realm and diversifying there? Why is the huge branch of the evolutionary tree that comprises insects so dramatically successful away from water—in the air, in forests and fields, in dung and deserts, on carcasses and on our clothing, on the bodies of plants and animals, and in the soil—whereas their evolutionary exploration of the sea is at best marginal? There are seaweed flies, salt-marsh and mangrove mosquitoes, midges, a variety of beetles and bugs, and even a few caddis flies, some of which parasitize sea stars as larvae; but there is a striking absence of dragonflies, damselflies, mayflies, stone flies, grasshoppers, crickets, cockroaches, termites, moths, wasps, and ants. As peripheral and unobtrusive members of marine communities, insects are not major players in the living theater of the sea.

On the surface, these observations constitute one more example of esoteric knowledge that may excite an ivory-tower scientist like me, but what relevance does it have for human affairs or even for a wider scientific public? As I see it, the most arcane curiosities often have unexpected implications, revealing general phenomena that may directly involve people.

In this particular case, the phenomena unmasked are colonization—of new habitats, distant geographic regions, or new economic roles—and incumbency, the power of the establishment to keep out intruders. Why have fish been unable to colonize the land when all the world's amphibians, reptiles, birds, and mammals are ultimately derived from an aquatic, fishlike ancestor? Why do dandelions, sunflowers, and thistles remain annual herbs in Europe and North America whereas some of their relatives have evolved into trees on the Galapagos and Canary Islands? How do economically subordinate organisms and human societies attain positions of economic dominance if those in power resist being displaced? By illuminating the conditions that prevent colonists or pretenders from establishing themselves in systems or positions foreign to them, we gain insights into the nature of invasion and into the vulnerability of entrenched incumbents. Moreover, the dissection of immigration reveals a curious parallel with climatic warming: in the short run, invasion or usurpation is usually harmful to the natives whose territory or status is being infringed, but over the long haul, they usher in a new day by erasing the damage and by establishing a new order.

But I am getting ahead of myself. Before tackling the larger issues, I turn again to the insects. Why don't we see insects on reefs, and why do so very few species live submerged in water? After all, some insect larvae and even some adults in freshwater have evolved means for retaining and using oxygen gas for long periods of time while they are immersed in water, so that they rarely have to come to the surface to replenish their supply of oxygen.

Possible answers come readily to mind—perhaps too readily. One possibility is that insects do not tolerate the high salinity of full-strength seawater, whose salt content is about thirty-five parts per thousand. But this is not correct. Not only are insects very abundant in inland salt lakes, but they have been found in brines with a salt content of 118 parts per thousand. Some groups—caddis flies and even some damselflies—are much more diverse in inland saline lakes than in the sea.

Oxygen concentrations in freshwater are 25 to 30 percent higher than in the sea, and rise by a factor of 1.6 as the temperature of freshwater drops from 68 degrees Fahrenheit (20 degrees Celsius) to the freezing point. These facts raise the possibility that submerged insects cannot maintain free oxygen in the body as easily in the sea as they can in lakes and streams, and that the transition from life on land to life immersed in water would be most difficult in hot climates. These prohibitions are, however, far from absolute. Some midge larvae that are restricted to the tidal zone in the North Sea can descend to a depth of ten meters or more in the brackish Baltic Sea. Many insects, including bugs, beetles, and water striders, have made the transition to marine life in the tropics.

Perhaps insects cannot take the heavy surf on exposed rocky seashores. This, too, seems unlikely, for many insects are well adapted to torrential flows in freshwater streams, and others live on the wave-swept shores of the American Great Lakes. Moreover, the small size of most insects would make them ideal candidates for life in environments with strong water motion. Their small bodies are so close to the surfaces on which they cling or crawl that flow in this so-called boundary layer is greatly reduced. Wave-battered rocky shores and the exposed stones of fast-flowing streams are, in fact, often covered with minute animals like snails, which simply do not experience the intense physical commotion that larger organisms like us encounter in such environments. Even if insects were unable to adapt to great turbulence, they should have no problems in the many calm environments that the sea has to offer. Yet even in these habitats, insects are a tiny minority.

Many insects live within, feed on, or pollinate flowering plants. Only sixty of some three hundred thousand living flowering-plant species are habitually immersed in seawater, and another few hundred species tolerate partial submergence in salt marshes and mangrove swamps. None of the fully immersed species is pollinated by animals. Marine insects are therefore effectively excluded from ways of life that on land have made insects by far the most species-rich group of living animals. In principle, however, marine insects could consume sea grasses and seaweeds; and they could even help "pollinate" sedentary marine plants and animals by ferrying sperm from one individual to the eggs of another. As far as I know, no instance of such animal-assisted cross-fertilization has ever been described for any marine species. Whatever limitation exists on mobile animals helping sedentary marine creatures to mate, it applies not only to insects but to all other potential vectors as well.

To be sure, the medium of water differs dramatically from the medium of air, meaning that the adaptations of water-dwelling insects are unlikely to work well on land and vice versa. Water at 68 degrees Fahrenheit (20 degrees Celsius) is some sixty times more viscous than air, meaning that locomotion for a water-dwelling insect is vastly more expensive than for a land dweller. Yet gravity is a major force on land and a negligible one in water. An animal on land weighs five to fifty times more than it does in water. Oxygen and carbon dioxide diffuse into and out of the body at a rate ten thousand times faster in air than they do in water, and the concentration of oxygen in water is only about 5 percent of that in air. Because small insects depend on diffusion to obtain oxygen, water dwellers need a much greater surface area through which oxygen can enter than is the case among land dwellers. This limitation effectively eliminates insects with a high metabolic rate—that is, a high oxygen demand—from aquatic habitats. Slow diffusion in water would also severely restrict communication by pheromones or other substances that are detectable at a distance, although predatory snails, sea stars, and crustaceans use long-distance chemoreception extensively to detect prey and enemies.

These and other proposed explanations are specific to insects. They all fail because the physical limitations of insects in saltwater environments either do not exist or have been evolutionarily circumvented by one lineage or another. Perhaps, then, we are asking the wrong question. Instead of seeking answers that focus on the physical capabilities of insects, we should look to the marine environment itself, or to the border between land and sea or between fresh and salt water, across which all insects must pass if they are to colonize the sea evolutionarily. Furthermore, the problem is by no means limited to insects. Many other groups of land-dwelling organisms have rarely or never entered the sea. There are no marine mosses, ferns, conifers, rodents, bats, amphibians, marine-derived members of the major land-snail groups, or sea-going songbirds; and the few spiders, scorpions, millipedes, centipedes, and mites that have entered the seas are minor players at best in marine habitats. Very few marine lineages have successfully adapted to conditions on land. There are no terrestrial echinoderms—sea stars, sea urchins, and their relatives—or cephalopods, and no lineages of brown or red seaweeds have ventured on to the land. In other words, insects are symptomatic of a much broader phenomenon, whose explanation must correspondingly apply to many lineages and many evolutionary transitions between physically contrasting environments.

I had long been acquainted with the sharp divide between land and sea, for it was on the bare rock surfaces of seashores around the world that I conducted research for my doctoral dissertation. I had become interested in how snails of marine ancestry adapted to the unrelentingly harsh conditions in this environment. At low tide, high-shore snails are exposed to hot sun, drying winds, and drenching rains. Food is not only sparse, but available mainly when snails can move about actively during high tide or while being wetted by spray from breaking waves. All of the approximately twenty-five snail lineages that have become specialized for life on the high shore have adopted a passive way of life, enduring periods of aerial exposure by suspending activity. It is an effective strategy for dealing with chronic inclemency, but not one that is well suited for crossing the border into a realm where the absence of salt water is compounded by the presence of active animals who are already highly adapted to life on land. Passive snails, or for that matter almost all other colonists from the sea, can in principle eke out a marginal existence in the midst of a thriving community of life on land, but they stand little chance of competing effectively with organisms that are already able to feed, grow, and reproduce effectively out of water. In other words, the high shore is an unforgiving filter, in which only passive species can exist; and well-adapted incumbent species on land prevent most potential marine colonists from evolutionarily immigrating to the dry land.

Evolutionary colonization of the sea from the land is equally problematic for the same reasons. Almost all marine insects cease activity when submerged in seawater. They cannot eat, mate, use force to defend themselves, or do much of anything else while their potential competitors—crustaceans, worms, snails, sea urchins, sea stars, fish, and other animals that are well adapted to the marine environment—go about the business of life. It is therefore not only the new challenges of the physical environment, but even more the biological environment that offers resistance to invaders from other regimes. Well-adapted incumbents enforce borders between realms of life and make these borders harder to cross than would be the case if the target environment were to be unoccupied.

If a lineage is to transcend a marginal existence in the new habitat, it must evolve means of staying active. Organisms venturing onto the land must be able to take up and eliminate gases in air instead of by absorbing and releasing them in water. They must overcome the much greater gravitational forces to which a body in air is exposed: animals and plants weigh thirty to fifty times as much in air as they do in water. Full independence from water requires that land organisms create an internal environment where egg and sperm can unite, as in a seed or through copulation. Once these adaptive accommodations to the land have been achieved, a return to water can be difficult, because the traits that make active life on land possible impose barriers to activity under water. It is thus the inactivity in the new habitat—the absence of economic engagement with incumbent competitors—that prevents most would-be colonists from successful invasion as major players.

The best opportunity for making the great ecological leaps between water and land, or for that matter between land and air, occurs when potential immigrants encounter little or no competition in their new surroundings. For example, formidable competitors roam the land today, but that was not the case more than 475 million years ago. At that time, the dry land would have been clothed in at most a microbial or lichen cover, for there were no land plants as we know them. If animals were present—evidence for them is at best circumstantial, consisting of possible burrows—they would have been small creatures feeding on decaying organisms. Even quite passive refugees from the competitive arena of the seashore could have established themselves successfully on land under these physically rigorous but biologically permissive conditions.

With the exception of pulmonate (lung-breathing) land snails, which probably came from marine ancestors some 150 million years ago during the Late Jurassic period, all the major groups of land animals had adapted to the land no later than the Early Carboniferous period, about 340 million years ago. Most colonizations from sea to land thereafter involved minor groups of snails and crustaceans—some crabs, pill bugs, sand fleas, and a few hermit crabs—but they never achieved the diversity and ecological prominence of the spiders, insects, mites, and four-footed vertebrates whose ancestors arrived early. The most successful colonization from water to land thus occurred at times when competition from terrestrial incumbents was minimal.

A few interesting cases of terrestrial invasion have occurred on isolated oceanic islands, where land-dwelling incumbents may offer less resistance. The most spectacular example is the coconut crab—a large, powerful hermit crab whose larvae disperse in seawater but whose adults live on land without a shell. Coconut crabs—so named because they are reputed to be capable of breaking coconuts—are found widely on islands in the Pacific and Indian Oceans but never on surrounding continents or on the large continental islands of Indonesia, New Guinea, and Australia. I suspect that the highly vulnerable soft stage of this animal as it molts would make it easy prey in land areas where mammals are important predators. With the exception of bats, mammals rarely reach islands without human assistance. Native island predators, including birds, may not pose as much danger to unprotected prey that rats, raccoons, mongooses, and a host of other mammals do on larger land masses.

Incumbency also controlled the transition from water to dry land in plants. Despite the considerable number of evolutionarily distant groups of marine plantlike organisms—red, brown, and green algae, as well as many single-celled lineages—only the green algae, among which are the ancestors of today's land plants, gained a significant foothold in habitats away from water. The oldest plants possessing water-conducting and food-conducting tissues in the stem were liverworts, hornworts, mosses, and an assortment of leafless and rootless forms that superficially resemble members of the desert community today. Roots, leaves, and seeds evolved independently in several plant lineages during the Devonian period between 420 and 370 million years ago. By the middle of that period there were forest trees, and by its end we see the first seeds, indicating the evolution of fertilization in the absence of water. Once the great group of land plants appeared in the Silurian period or perhaps even in Ordovician time, as early as 470 million years ago, no other lineage of water "plants," or algae, made the transition to life on land.

Transitions between sea and land are rare, but many lineages have entered freshwater lakes and streams from the land. At least two hundred lineages of land plants have colonized freshwater, where about a dozen of them have taken up the unusual habit of floating on the surface without being rooted in the bottom. Only three lineages of flowering plants have evolved to live in the ocean, and surface floaters are absent there. By my conservative estimates, there are at least twenty-four mammalian lineages with freshwater members, whereas only seven have entered the sea. Marine-to-freshwater transitions have also been frequent: there are at least fourteen in clams, thirty-three to thirty-eight in snails, two or more in crabs, and hundreds in fish, including sharks, rays, herring, cod, sticklebacks, gobies, and salmon. Seals and dolphins have secondarily colonized freshwater from the sea on several occasions each. Many of these transitions are geologically recent, and there are indications that some sticklebacks and small crustaceans have entered coastal freshwater lakes in North America and Europe repeatedly within the past ten thousand years. Freshwater habitats tend to be small and ephemeral compared to marine and terrestrial ones, especially in regions that were once covered by ice during the Pleistocene epoch. They are thus like biological islands in that the sophistication of competitors and predators—and thus of incumbents—is modest. And yet, barriers remain. There are no freshwater echinoderms (sea stars, sea urchins, sea cucumbers, and their relatives), cephalopods (squid and octopus), brachiopods (lamp shells), corals and their allies, barnacles, land-derived snails, social insects, or fossil trilobites. The factors keeping these organisms, many of which would have been formidable competitors and predators, from colonizing freshwater are unknown. Barriers, it seems, are never fully dismantled.

Two groups of organisms—vertebrates and to a lesser extent land plants—have on several occasions given rise to marine lineages that have not only competed effectively with well-established incumbents, but have in some cases attained dominant positions in the sea. Whales and seals are among the top predators in the ocean, and sea cows (manatees and dugongs) are major herbivores. Sea grasses are among the most productive marine plants. What is the secret of success of these land-derived groups? What can they do that most insects, spiders, scorpions, land-based crustaceans, centipedes, millipedes, and mosses are unable to accomplish in the marine environment?

In contrast to insects, land-based vertebrates that evolve marine habits remain active during the transition, and therefore are effective competitors with native marine life from the earliest phases of occupation. For example, the earliest whales, living some fifty-two million years ago during the Early Eocene epoch, were probably amphibious carnivores. Over the next several million years, these animals eventually evolved into adept swimmers, achieving a nearly worldwide distribution. The earliest sea cows, which entered the sea at about the same time as the whales, may have been amphibious plant-eaters. Similar paths to marine life were taken by mosasaur lizards beginning about ninety-five million years ago, by the ancestors of seals about twenty-three million years ago, and by an extinct lineage of sloths in Peru that lived between eight to four million years ago. Reptiles—several lineages of turtles, crocodiles, lizards, snakes, and fossil groups, including plesiosaurs and ichthyosaurs—evolved marine habits at least thirty times beginning 247 million years ago during the Early Triassic period, very shortly after the end-Permian disaster.

Because they nest on land (or, in the case of jacanas, on floating freshwater vegetation), birds have never fully severed their connection to the terrestrial realm, but many lineages have given rise to birds that feed extensively or exclusively on marine shores or in the open sea. Such ecological transitions have occurred in at least six major lineages, including grebes and their relatives, the waterbirds (penguins, petrels, pelicans, and the like), shorebirds (gulls, terns, plovers, sandpipers, and auks), ducks and geese, sea eagles, and two or three extinct groups. Already during the Cretaceous, in the age of dinosaurs, marine birds were important consumers, first as predators, and later—with the evolution of geese and some lineages of ducks—as herbivores. Moreover, the excrement that these birds leave on land in the form of guano is in many parts of the world an important source of nutrition for plants, and thus constitutes a key subsidy of the land by the sea. Through their travels, birds as well as seals have forged a key nutritional link between land and sea, and contributed importantly to the selective environments of both realms.

All these vertebrates became important consumers in marine ecosystems, and several—killer whales, sperm whales, elephant seals, and

Mesozoic ichthyosaurs and crocodiles—became top predators there. Not only did the less fully marine ancestors of these animals penetrate seas that were already well endowed with incumbents, but they gave rise to descendants that equaled or even surpassed the capabilities of such native marine groups as sharks, reef fish, tunas, and giant squid, which evolved alongside them.

The only other land-derived organisms that met with great ecological success in the sea are plants. Salt-tolerant mangrovelike trees may have existed in periodically submerged saltwater environments as early as the Late Carboniferous, about three hundred million years ago; but the three lineages of sea grasses, and the thirty or so other lineages of plants that can tolerate temporary immersion in tidal waters, all belong to the great evolutionary group of flowering plants, which occupied nearshore environments beginning eighty million years ago. What makes these invaders unusual is that they did not, for the most part, compete directly with incumbent seaweeds. Two important characteristics set the land-derived plants apart from the red, brown, and green seaweeds. First, the new arrivals have roots, which enable them to tap raw materials from the sand and mud in which they grow. Except for a few unusual green seaweeds, which have creeping rootlike structures that are capable of drawing nutrients from the sand in which these algae grow, most seaweeds attach to hard objects by means of a holdfast, a specialized structure that fixes the organism to the seafloor but that does not aid in obtaining nutrients. Most seaweeds therefore depend on water-borne substances rather than on the rich source of food in mud and sand. This limitation restricts seaweeds to agitated waters, where nutrients are quickly replaced as they are taken up by the body of the plant. Seaweeds have also generally not colonized sandy and muddy environments, where marine-adapted flowering plants are effectively the only significant sources of oxygen and fixed carbon. Flowering plants have thus come to dominate these shores, where they form extensive seagrass meadows, salt-marsh vegetation, and mangrove forests. In effect, they have created a way of marine life that barely existed before flowering plants evolved some 140 million years ago during the earliest Cretaceous period. In their new surroundings, these flowering plants have vastly increased the productivity of nearshore environments and thus stimulated both the evolution of native marine lineages and the invasion of vertebrates from the land.

If the absence or removal of well-adapted incumbents were solely responsible for the opportunistic entry of land dwellers or freshwater species to the sea, or of marine species onto the dry land, the great mass extinctions of the geological past should have triggered many such border-crossing episodes. Incumbents holding competitively dominant positions have proven to be highly vulnerable to extinction during global crises. The power that these creatures wield is contingent upon high productivity, but if production is severely curtailed as the result of a collision between Earth and a celestial body or of enormous volcanic outbursts that darken the sky and prevent sunlight from reaching photosynthesizing plants and plankton, the demand cannot be satisfied, and populations of dominant animals become unsustainable. Studies of fossil mammals by UCLA's Blaire Van Valkenburgh and others show that top carnivores, which are dominant members of their terrestrial communities, are replaced far more often by subordinate lineages than are herbivores and smaller carnivores. The large body size of top carnivores—and of such herbivores as elephants and dinosaurs—offers advantages in competition against predators, and in reproduction, but it is also associated with small population size, which in turn makes populations vulnerable to random fluctuations in climate. A small population does not have far to fall before it becomes too small to be sustainable. Given the long-term susceptibility of large-bodied dominants to climate-driven variations in productivity, mass extinctions and the demise of the ruling hegemony should offer unusually favorable circumstances for colonization of imperfect immigrants from alien environments.

However, for the most part this does not seem to be the case. Evolutionary transitions between sea and land generally took place millions of years after mass extinction events, at times when the climate warmed and populations of many species expanded. Mosasaur lizards entered the sea before, not after, the moderate extinction events that occurred ninety-three million years ago. The first whales and sea cows appeared thirteen to fifteen million years after the great end-Cretaceous extinction, and seals entered the sea about twenty-three million years ago, some ten million years after the destabilizing events at the transition between the Eocene and Oligocene epochs. Turtles entered the sea about 220 million years ago during the Late Triassic, long after the end-Permian extinction, and again during the Jurassic and Early Cretaceous, again far from the times of crisis.

The single exception is the time immediately following the greatest of known mass extinctions—the catastrophe at the end of the Permian period and Paleozoic era, about 251 million years ago. In the recovery from this event, at least seven distinct lineages of land-dwelling vertebrates gave rise to marine species during the first five to seven million years of the Early Triassic period. Many of these early marine reptiles were slow, rather heavily armored swimmers whose teeth were well suited for crushing hard-bodied prey. Ecosystems of the Early Triassic seem to have been devastated in ways that have not been repeated after other mass extinctions. Judging from the very small size of many Early Triassic marine animals—snails had shells at most seventy mm (two and a half inches) long—productivity must have been very low. There is also evidence from carbon isotopes that have been recovered from sediments to indicate great fluctuations in the global ocean's cycling of carbon through the system. Stable relationships among species had therefore not been reestablished. Fast predatory fish that are capable of catching and killing slow newcomers coming from the land appear to have been absent from Early Triassic seas. No similar large-scale invasion from the land took place after the end-Triassic crisis, about two hundred million years ago, or in the epoch following the mass extinction at the end of the Cretaceous period, about sixty-five million years ago. The fossil record indicates that several fast predators, including sharks, bony fishes, and (after the Triassic) marine reptiles survived the crises and might therefore have offered significant resistance to the establishment of clumsy swimmers.

To me, the most significant finding about marine colonization from the land is that successful instances occur during times of global warming, when diversity in both the donor and recipient environments is rising and resources are plentiful, especially in the sea. The first marine mosasaurs, whales, sea cows, turtles, and sea grasses became established during conspicuously warm geological intervals. Seals, sloths, and sea otters went marine on highly productive, relatively warm coastlines where the native flora and fauna were already highly diverse. It could thus be that, although incumbents were surely present during these invasions, the ecosystem during times of expansion is sufficiently forgiving to accept active immigrants. There is room for imperfect newcomers to establish themselves and ample opportunity for these species to adapt.

If mass extinctions often fail to create opportunities for lineages in one environment to take the place of extinguished incumbents in another, is the whole hypothesis of incumbency—the idea that the elimination of the powerful elite creates evolutionary opportunities and subordinate survivors—incorrect? I do not think so. The reason is that, despite great losses during times of crisis, many incumbent lineages survive. Even a competitively subordinate marine lineage is better adapted to the sea than are most potential colonists from the land. Vanished marine incumbents will therefore generally be replaced not by terrestrial immigrants, but by lineages invading from other parts of the marine realm. Likewise, losses on land are made up by colonists from other continents, not from marine lineages venturing onto dry ground. The physiological hurdles that must be crossed at the border between sea and land remain high.

Transitions between such radically different environments as land and sea represent extreme examples of incumbents keeping out would-be colonists, but the resistance that native species individually and collectively offer to immigrants also figures prominently in patterns of interoceanic or intercontinental migration. The recipient regions that were discussed in chapter 10 are not only smaller and often less productive than the donor regions from which many invaders come, but usually also more heavily disrupted and therefore less resistant to colonization. When North Pacific lineages reached the North Atlantic by way of an ice-free Arctic Ocean beginning about three-and-a-half million years ago, they encountered a fauna that had been ravaged by extinctions. Even the successful invasion of islands by human-introduced species from the continents is greatly facilitated by habitat destruction and human-caused extinctions of native island species. Offshore islands near New Zealand that have not been disturbed by humans have been far more successful in repelling alien species than islands that have been deforested and where deer were deliberately introduced.

Most of the ecosystems that will be left as we continue to expand our hold over the biosphere will have been so diminished in size and in the abundance of resistant incumbents that they will be the target of increased colonization by foreign species. This does not bode well for keeping intact those remnants of forest, reef, or prairie that remain. As a conservative natural historian who revels in the distinctive species and relationships in ecosystems from different parts of the world, I deplore this heightened vulnerability to foreign invasion and the homogenization of nature. As a paleontologist, however, I understand that a changing of the guard is inevitable. Extinction, invasion, and accommodation have characterized ecosystems throughout history; and despite the myriad great and small disruptions that have tinkered with and sometimes dismantled ecosystems that were stable for millions of years, systems recover, reconstitute themselves, and prosper anew. Invasion by outsiders plays a large role in this renewal. In the end, it is important to preserve that regenerative capacity, even in those ecosystems that have been cobbled together by invaders coming from all corners of the world.

How do these ideas apply to human colonization and conquest? Historians have long understood that societies weakened by disease, famine, or war are particularly vulnerable to subjugation by invaders. Smallpox brought by early European visitors decimated native North American populations and hastened French, English, and Spanish colonization and expansion on the continent. Destabilizing conflicts in both Mexico and Peru contributed to the crumbling resistance of the Aztec and Inca empires against Spanish conquistadors. East Asian societies, by contrast, remained largely intact, and kept Europeans restricted to coastal trading forts. Strong group identity—a powerful aspect of incumbency in our species—enabled Russia to repulse the Napoleonic invasion and helped Vietnam defeat the militarily much more powerful United States.

In the prehuman past, most of the disruptions that placed entrenched incumbents at risk of being supplanted originated outside the biosphere. Geological convulsions in the Earth, collisions between Earth and celestial objects, and the ripple effects that these larger-than-life agencies had on the capacities of organisms to make a living were responsible for the great mass extinctions and for the innumerable lesser calamities that are so well revealed by the fossil record. What makes the human situation unusual is that many of the disruptions to civilization are caused by our own species. Terrorists, financial speculators, the invention of weapons of mass destruction, overexploitation of resources, and pollution belong to us. It is our decisions, actions, and values—both good and bad—that are the chief agencies of change. We can cause irreparable harm, but we can also create great good.

When looked at over the short-time scale of individual lifetimes, the disruptions that force incumbents out of office are calamitous. The removal of top producers and top consumers leads to a decline in economic vigor, a recession of sorts in which supply and demand are diminished. Over the long sweep of geological time, however, evolution replaces the old guard with new leaders from the ranks of previously subordinate lineages; and as these new top competitors penetrate nearby regions and ecosystems, a new order is established, often one in which the standards of performance eventually surpass those in the old regime. Full replacement, such as that of Cretaceous dinosaurs by large Cenozoic mammals and birds, or of seagoing Mesozoic reptiles by whales, may take millions of years, but the capacity for renewal and for going beyond the previous status quo is ever-present as lineages evolutionarily vie for power. Similarly in the human domain, the demise of empires and of ruling elites is followed sooner or later by the rise of new ones. The old order may not be surpassed for centuries—think of the fall of Rome and the birth of a technologically more advanced civilization in Italy more than seven hundred years later—and whether the new civilization is "better" than the old is subject to interpretation—but the point is that catastrophic social disintegration eventually is succeeded by a different civilization. As in the case of global warming, discussed in chapter 9, the destructive short-term effects of change lay the groundwork for long-term opportunity to create a new order through adaptation.

In the context of incumbency and adaptation, I can rightly be accused of being a relativist, one who holds that success depends more on comparative capacities, as measured against those of competitors, than on absolute standards of performance. An adaptation cannot be judged on its own merits; it must instead be judged according to how well it works in competition with other ways of coping with prevailing challenges. A land-dwelling beetle and a freshwater mosquito larva may still be able to make a living in a few restricted coastal settings where competition from incumbents is minimal, but not in marine environments where better-adapted animals dwell. A civilization such that of the Aztec, Maya, or Inca, which lacked metal weapons, domesticated-animal labor, and wheeled transport, was adequate for its time and place, but not when Europeans came over the horizon in ships with horses, cannons, diseases, carts, and a unifying religion. Hunting and gathering food worked well for populations of hominids during the Pleistocene, but became confined to a few economically marginalized societies with the spread of farming. In short, the standards an organism or a society must meet in order to survive and prosper evolve. If adaptations do not change to meet the new standards, their bearers must find places where the criteria for success are commensurate.

Incumbents rise and fall, powerful elites succeed one another, and colonists expand into new areas. They all take part in history. They are not only swept along as circumstances unfold, but they contribute to their fate. Is history simply a long litany of events linked by chains of cause and effect, or does it reveal patterns discernible over very long stretches of time? What determines the times and places where great historical events take place? I explore the nature of history in the next chapter.

CHAPTER TWELVE

The Arrow of Time and the Struggle for Life

We are all products of the past. History has left an indelible mark on our anatomy, our physiology, and our cultures. We can describe, reconstruct, and explain pathways of evolution, patterns of descent and marriage in human family trees, the origins and outcomes of battles and wars, the economic relationships and performance levels of fossil organisms and their parts, and sequences of climatic and geological events on the Earth. But is there anything to history besides narrative? Does adaptive evolution hold implications for the future? Is there something emergent about the directions of evolutionary history?

To many observers, history is little more than an endless parade of actors and events, motives and circumstances, troubles and trajectories, going nowhere in particular but somehow ending up in the present. The passage of time ensures that everything is linked to some antecedent state. There is pervasive selection among organisms, and there are goals and aspirations that influence events in human lives and societies; but on the largest scales of space and time, there is no target end state, no supernatural force directing the sequences or choosing the pathways of change toward some prescribed outcome.

Individuals, ecosystems, and empires rise and fall; rulers replace one another—by inheritance, election, coup, or catastrophe—and there appear to be defining moments—discoveries, inventions, battles, and mass upheavals—that in retrospect bias the courses of history. Sequences begin and end with particular, contingent events and participants, which can be understood after the fact but which cannot be anticipated or predicted. There is only the passage of time, and even that cannot be taken for granted if the physicists are right.

In this contingent view of history, any narrative—whether it be a body's embryonic development, a person's biography, a nation's political journey, the evolution of life, or the unfolding of the universe—can be cast as a scientific description, interpretable in terms of causes, consequences, and feedbacks stemming from interactions. But change—or the lack of it—reflects the particulars of time and place, of circumstances intersecting with matter in motion. As Stephen Jay Gould puts the case toward the end of his vast book, *The Structure of Evolutionary Theory,* "As antecedent states are, by themselves, particulars of history rather than necessary expectations of law, . . . we regard these subsequent outcomes as unpredictable in principle."[1]

If chance were the chief agency of history, as Gould and others maintain, future states even one step from the present could never be foretold. Anything could happen at any time and in any place, without constraint and with no apparent order. Entities from molecules and cells to organisms and groups interact, combine, multiply, divide, and disappear in so many ways that the number of possible outcomes of their activities vastly exceeds the number of states actually realized. In this sense of contingency as randomness and particularity, contingency—the dependence of historical events on singular prior conditions that are themselves consequences of still earlier unique circumstances—must therefore be a universal property of all systems, living as well as nonliving. In the view of many modern thinkers, including Gould and Stuart Kauffman, the universe in general and the domain of life in particular are nonergodic—that is, nonrepeating: the course of history at one place and time will not resemble that in another. Each event, participant, circumstance, sequence, and pathway is unique. According to this view, the likelihood of realizing the same outcome more than once from similar prior conditions is vanishingly small. Even if we could fully describe the immediate past, we would still be unable to guess what comes next. The future cannot be known until it happens.

But this view of history is incomplete. The claim that history is nothing more than random sequences without repetition, predictability, or emergent direction conflicts with both fact and theory. Life, I argue, creates its own future and its own predictability through adaptation, and imparts to history an arrow of time, an overall direction emerging from selection and competition despite the background noises of contingency. Winners—agents that by virtue of their adaptations gain more or lose less than their adversaries—can attain power only in a limited number of ways, which are achieved independently in multiple evolutionary lineages. The surviving losers likewise have few adaptive options. The details of the adapted state or the pathway toward it will vary endlessly, and thus remain highly contingent and particular, but as long as power is rewarded and mechanisms exist to increase it or adapt to it through selection and through positive feedbacks with resources, there will be a general upward trajectory toward greater power, control, and predictability in the biosphere. This upward trajectory is the arrow of time in the history of life.

A recurrent theme of this book is that adaptation represents a way of extracting predictable signals from initial chaos and then responding accordingly. The existence of evolutionary lineages is one expression of this process. We would be unable to infer evolution if genes and the traits they express were not transmitted from generation to generation—that is, if there were not some predictable signal through time. Some lineages, in fact, are so conservative—and thus so predictable—that ancient members can hardly be distinguished from modern ones. Their adaptations work well in a given environment, and if individuals can track that environment through time, they can persist for eons. The lamp shell *Lingula,* for example, leads a low-energy existence in burrows on warm mudflats. Their simple shells can be traced back with only minor variations to the Cambrian period, more than five hundred million years ago. The realized range of possibilities for lineages like this one is extremely limited, and the future will be like the past. For members of these conservative groups, the world is all too predictable.

Adaptation—one of life's emergent properties—exposes patterns, or repeated sequences and correspondences, that are detectable by even the simplest organisms. The patterns that matter are the opportunities and dangers influencing survival and the ability to multiply. For inanimate objects, the number of possible states that are one step beyond the present may be infinite, but for living things this number is quite small, because most of the potential states are unworkable and therefore unavailable. Prediction in the form of adaptation may be inductive—an extrapolation from the known past to the unknown future—but its validity is verified whenever the adapted organism passes the test of selection. As I noted in chapter 4, adaptation lessens the consequences of unpredictability and extends the range of phenomena that can be predicted. It therefore extracts and creates signals that render the courses of evolution less haphazard.

One telling indication of the limited number of adaptive options to be found is the repeated, independent acquisition of adaptive breakthroughs. Inspection of the tree of life shows that even very complex processes and states have arisen multiple times, often from radically different starting points. Photosynthesis evolved at least three times. Lignins—the polymers that stiffen plant-cell walls and that enable land plants to grow erect stems in the face of gravity—evolved both in the ancestors of vascular land plants and in some marine red algae. The so-called eusocial condition—the social organization with specialized castes found in ants, termites, snapping shrimps, and naked mole rats, among other animals—arose at least nineteen times. Warm-blooded physiology was attained to varying degrees in at least eight lineages of vertebrates and in many insect groups. Rootless, free-floating water plants evolved from rooted ancestors on twelve or more occasions. Mineralized skeletons, such as snail shells and barnacle tests, arose in no fewer than twenty-one algal and animal lineages between 550 and 445 million years ago and on several occasions since. Ferns and the ancestors of flowering plants evolved a closed network of leaf veins more than fifty times from an openly branched system. Venom injection evolved in a minimum of twenty-four animal groups, ranging from jellyfish and their relatives to spiders, scorpions, centipedes, stinging Hymenoptera (ants, wasps, and bees), octopuses, cone snails, scorpion fish, fangtooth blennies, snakes, and two or three mammals including the Australian platypus and the North American short-tailed shrew. Feathers are unique to birds and, in simplified form, their theropod dinosaur forebears, but they belong to a class of insulating structures that also includes hair in mammals and the fuzzy coat of warm-blooded bumblebees.

Some biologists question whether these separate adaptive acquisitions are really independent, because the genes required to build the relevant structures and physiological controls were already in place in ancestors that did not have the adaptations. The genes that express light-sensitive proteins, which are essential for the evolution of image-forming eyes, occur in bacteria as well as in animals that perceive light but lack eyes. The capacity to form eyes could thus be construed as having evolved only once in some very remote ancestor in which the light-sensitive proteins first appeared. I would argue, however, that the mere existence of enabling genes does not guarantee any adaptive outcome. The genes responsible for jointed arthropod limbs, for example, also construct unsegmented horns used by many beetles for combat and sexual display. A pile of bricks may be necessary for building a house, but it could just as easily become a paved road or a decorative wall. Potential must exist, but there must be selection to turn it into functional reality.

Just as compelling are the many independent "experiments" with similar outcomes, in which related but geographically separate lineages evolve in parallel at different times, on separate continents or islands, or in mutually isolated bodies of water. Predatory vertebrates with long, sharp-edged, laterally flattened saberlike upper canine teeth evolved on many occasions, first among some Late Permian therapsids ("mammal-like reptiles"), and much later among Australian marsupials and several North American and Eurasian lineages of cats, all now extinct. Among West Indian *Anolis* lizards, tree-dwelling species occur on each of the larger islands. These arboreal species all look alike, and were once thought to have a single tree-dwelling ancestor that spread throughout the Caribbean. However, molecular work by Harvard's Jonathan Losos and his colleagues has definitively shown that the tree-dwelling habit evolved separately and in parallel from separate ground-dwelling ancestors on each island.

Although all of these repeated adaptive states must have had a unique first occurrence, I cannot think of a single adaptation whose attainment depended on a unique historical event, transition, or evolutionary point of departure. Adaptive breakthroughs within lineages, and new partnerships among lineages may be rare, but they have a very high probability of taking place given the passage of enough time. In the words of the Belgian molecular biologist and Nobel laureate Christian de Duve, "[E]volutionary pathways may often have been close to obligatory, given certain environmental conditions, rather than contingent and unrepeatable, as received wisdom contends."[2] Moreover, new pathways lift constraints and permit some emerging lines to evolve further along adaptive directions, raising the performance standards of other lineages and imparting an overall, if diffuse, direction toward greater adaptive freedom and versatility.

Predictability, however, does not automatically produce an arrow of time. Organisms can discern and create a pattern by formulating an adaptive hypothesis of their environment, but how does the collective "wisdom" represented by those adaptations yield emergent directionality? How can the contingencies imposed by prior states be overcome and undone? What are the circumstances—external conditions as well as life's economic activity—that release evolutionary lineages from the tyranny of origin? The answers reside in how opportunity is created, and more important still, in how living things capitalize on those opportunities.

To breathe some life into such heady matters, I draw on my own life history. When I listen to the music of the great Renaissance and Baroque composers, I often contemplate how my life would have unfolded had I been born during the sixteenth or seventeenth centuries when this marvelous acoustic art was being created. For someone of my station in childhood—blind and living in a poor if loving and devoted family—those beginnings would likely have meant a life of illiteracy, poverty, and ignorance. Society, then, had neither the wealth nor the will to create the opportunities that would loosen the yoke of early circumstance, nor for the most part did it sanction ideas that ran counter to received religious orthodoxy. If I had had the cast of mind I do today, my life would have ended on the stake or the scaffold.

But I was raised in societies that not only possessed the resources and technology to overcome individual shortcomings, but also embraced the goal of spreading opportunity to everyone. The nineteenth-century Frenchman Louis Braille, taking advantage of our exquisitely sensitive hands that we inherited from our primate ancestors, invented a tactile script that enabled the blind to read and write easily for the first time. Thanks to government-sponsored institutions—schools, agencies for the blind, and foundations for financing scientific research—as well as to the activism of social reformers and the benevolence of philanthropists, a motivated blind person from a supportive family could receive a first-rate education, secure a satisfying position in society, and enter the life of a scientific scholar in settings where freedom of thought is for the most part revered. The constraints of origin that would have set the course of my life centuries ago in accordance with prevailing social and technological norms had given way in my actual life to a far more permissive environment in which many more options became available.

This rather self-indulgent case illustrates how the constraining influence of prior states can be partially blunted. Three circumstances would seem to be required: (1) a structural organization conducive to innovation, or in my case the existence of educational and scientific institutions promoting free inquiry; (2) sufficient resources—in other words, wealth—allowing costly improvements such as educating disadvantaged segments of the population to be made without unduly compromising existing functions, and (3) intense selection (or, in the case of human lives, strong will and ambition, instilled and encouraged by society at large) in favor of high performance. Given sufficient time, these circumstances and the feedbacks among them can lift the burden of the past and usher in an era of greater freedom and adaptation.

The organizational structure or body plan that is most conducive to adaptive modification and to the introduction of new capacities is an arrangement of several component parts, or modules, each dedicated to a specialized function. The parts operate semiautonomously yet cooperatively, and there is enough redundancy among them that, if one part should fail, the whole body can continue to function.

The parts are flexibly controlled by a central authority, a brain or, in the case of human institutions, a governing body. An organism or group with such an organizational structure is able to perform many tasks and provides evolutionary versatility, the potential to "explore" more than one adaptive pathway.

To gauge the adaptive versatility of the body plan of a lineage of organisms, we can count the number of functionally distinct modules—such as cell types, skeletal elements, body segments, and flower parts—and determine the number of regulatory genes that would be needed to account for the differentiation of these modules and the boundaries between the parts. In a bilaterally symmetrical animal, the right and left sides are essentially identical, so that just one parameter—and one gene—suffices for this aspect of the body's construction. In a body in which one side differs consistently from the other, a minimum of two parameters would be required. As more parameters—and more regulatory genes—are added, the range of potential adaptive options, or morphospace, is enlarged, providing more evolutionary degrees of freedom and enabling the lineage to depart more easily from its previous pathway of change. Greater versatility loosens the constraints imposed by the circumstances of origin.

I gained a grasp of these principles of versatility by playing with blocks. As a young faculty member with an interest in these matters, I whiled away countless hours in meetings by constructing different configurations with a fixed number of plastic cubes. With just two blocks in perfect face-to-face contact, only one configuration is possible; with three blocks, there are just two, and with four there are already five. By allowing edge-to-edge contact, by increasing the number of cubes, and by varying the sizes of cubes, the envelope of possibilities expands enormously. As it does so, it becomes increasingly feasible to construct approximations to desired functional shapes—tables, chairs, buildings, and the like.

Well understood processes push organisms toward greater architectural freedom. The first step is a mutation resulting in the duplication of a gene. Though redundant at first, the two copies can diverge with further mutation and selection, one daughter gene encoding the ancestral trait or function while the other takes on a new role. Duplication and subsequent differentiation yield a body plan with more domains, in which a given gene's expression becomes confined to a particular module or location without strongly affecting other domains of the body. The number of functions that the body can carry out also increases, expanding the body's performance capabilities.

The work of the gifted Dutch evolutionary biologist Frietson Galis and her colleagues on the vertebral bones of the neck illustrates this well. Nearly all mammals, ranging from rodents and primates to the long-necked giraffe, have just seven neck vertebrae, whereas birds have anywhere from nine to twenty-five, with the largest number being found in swans. Mutations in the number and form of neck vertebrae occur in humans and other mammals, but they are almost always lethal, because they affect many additional structures as well. Differentiation of the neck bones in mammals takes place during an early developmental stage, when each part of the body is developmentally tightly linked to every other part. If a mutation causes the formation of the neck bones to go awry, it will also affect the construction of other essential organs. In birds, by contrast, neck vertebrae differentiate at a late stage, so that when a mutation in their number or form occurs, that mutation has few side effects. Though still connected, the various regions of the body have become relatively more independent of one another, with the result that mutations affecting late events are less destructive, and can persist long enough, in many cases, for the altered structure to become part of a new adaptive architecture. Should the events of neck development be displaced to an earlier stage when gene expression is more global, the number of neck bones would likely stabilize, as has been the case in the vast majority of mammals.

In the evolutionary tree of plant and animal life, body plans with early origins near the tree's base or close to the ancestry of major branches possess few discrete modules, which are regulated by a limited number of regulatory genes and which therefore restrict these body plans to a small morphospace. Variations on the theme of these early foundations do not stray far from the ancestral type. Body plans evolving later, farther out on the branches, are more versatile, so that some of the evolved variations bear almost no resemblance to the founding lineage.

Early land plants, for example, grew in a highly stereotyped fashion with little differentiation of vegetative (that is, nonreproductive) parts. Repeated branching of the plant's main axis yielded two twigs of similar size at each node, or branching point. These twigs were simple cylindrical structures. As major new plant groups arose, the range of growth forms increased. In some, multiple branches issued from each node, and twigs were no longer constrained to be of equal length. In at least four separate plant groups beginning in the Devonian period, leaves came to be differentiated from twigs. At first, the leaf veins formed an openly branched pattern—that is, one in which the branches do not come together again once they have diverged, and grew only at the leaf margins. Open branching was replaced by a closed network of veins in at least fifty lineages, made possible by veins growing not only at the leaf edge, but also in zones of the leaf away from the margins. Still greater versatility was achieved when leaf growth was no longer just at the apex, or leaf tip, but also (or only) at the point where the leaf is joined to the rest of the plant. Although we know little as yet about the genetics of plant growth, I suspect that the more versatile body plans, such as that of flowering plants, are constructed with a larger number of regulatory genes.

In arthropods, the body is organized into segments, arrayed from front to back, with each segment bearing at least one pair of segmented appendages. Trilobites and other early arthropods have differentiated head segments, but their trunk segments all are alike in form and function. By contrast, advanced arthropods have elaborately differentiated limbs. Crabs, for example, have a pair of prehensile claws followed behind by walking legs and, near the posterior end, specialized appendages to which the female attaches her eggs. In some crab lineages, one of the pairs of walking legs is modified into paddle-shaped limbs by means of which the animal swims. Careful studies of all fossil and living crustacean lineages by Sarah Adamowicz and her colleagues reveal an astonishingly consistent trend through time toward greater architectural and functional differentiation and regionalization in the crab body as a whole and in the appendages. Work by Neil Shubin, Cliff Tabin, and Sean Carroll confirms that this increased versatility reflects a larger number of gene-regulatory domains. Adamowicz's work and my own investigations indicate that a pervasive trend toward greater versatility is observable in such additional major animal groups as molluscs, echinoderms, and vertebrates.

Importing genes or even whole genomes from other organisms clearly offers another avenue for innovation. As I noted in chapter 6, the usual way this occurs is by hybridization or through symbiosis. Reef corals did not acquire their ability to photosynthesize from scratch, through mutation of their genes; they did it by entering into an intimate partnership with dinoflagellates, single-celled planktonic organisms that could already photosynthesize. There is little possibility and even less incentive to reinvent the wheel—or the complex machinery for fixing carbon in organic compounds—when the relevant technology is already available for co-option.

In the end, evolution is not just about genes, but about when, where, and how genes are expressed. Although truly new genes occasionally arrive by way of symbiotic or parasitic guests or through hybridization, most adaptive evolution proceeds by modifying old genes or changing the activity, or meaning, of those genes. Alteration of the existing genome can proceed by mutation at a single nucleotide site, insertion or deletion of a segment of DNA, and rearrangement of genes along a stretch of genetic material; and some of these changes control when and where genes are turned on or off. In the words of Sean Carroll, the University of Wisconsin's eminent authority on genes and evolution, "Innovation . . . has been a matter of modifying existing structures and of teaching old genes new tricks."[3] In short, genes change much less than their functions.

For example, genes controlling outgrowth from the body in arthropods may encode instructions for limbs ranging from a spider's silk-spinning appendages to a lobster's mandibles, a blue crab's swimming limbs, a butterfly's wings, or an insect's head antennae, depending on where and how the genes are expressed. Within the same developing embryo, a given gene is associated with different structures depending on where and when it is rendered active or inactive by other genes. The boundaries of those domains of action are set by still other genes and by their joint interactions. The early embryo comprises a single domain, a module without internal genetic boundaries, but as development proceeds, it becomes subdivided into an increasing number of more local, semiautonomous spheres of genetic influence.

Human language offers striking and instructive parallels to the genomes of organisms, and serves as an accessible model of how functions, meanings, and the envelope of possibilities can expand by virtue of a flexible, compartmentalized organization. Its sequences mutate, borrow from other languages, and vary in the pattern of expression according to modifiers.

Like the nucleotide sequences of genes, the letter or sound sequences of words undergo slight alterations in composition and translation, but their meanings often change far more substantially, depending on the larger context in which they are expressed. Take the English word *knight.* It is derived from the same Germanic root that became the modern Dutch and German *knecht.* In its silent *k* and *gh,* the English word preserves orthographical vestiges of its ancestral pronunciation. The modern knight refers to a valorous horseman, whereas a *knecht* is a loyal but subservient workman. There are innumerable examples of duplication and differentiation, in which a given root gave rise to two or more words with subtly different spellings but divergent meanings. This is the likely origin of the English words *polite* and *polish,* imported to the language by French-speaking Normans but ultimately derived from the Latin root *polio.* In most languages, the same sequence of letters or sounds has different meanings depending on its context. Think of all the definitions of such common English words as *grave, lead, let, mean,* and *still.* The meanings of words, like the functions of genes, change when sequences combine. Prefixes and suffixes modify the functions of words; and two or more words with distinct meanings when used separately can be combined to form a new word with a new meaning. Words like *interrupted, fieldwork, substantially,* and *goatsucker*—to choose four at random—illustrate these points nicely.

Thus, as in genomes, emergent linguistic structures are created as words are strung together to form sentences, which form paragraphs, which come together as whole narratives. Syntax and grammar—the equivalents of genome-wide regulation—are components of the evolved glue that holds language together and enables it to function. In languages, as in genomes, there is a gradation in conservatism from letters or nucleotides to single words or genes to stories or organisms. The possibilities for modification, combination, complexity, and invention are open-ended and infinite. Neither words nor genes stand alone; they acquire function and meaning only through interaction in evolving contexts. Over time, languages are enriched and become capable of naming and describing a greater variety of objects, actions, and ideas as rules of usage become more flexible, new meanings and contexts arise, and sequences enter from other linguistic genomes.

Wonderful as the genome is for specifying instructions and creating complexity, execution of its full potential requires that the constructed bodies be nourished by abundant and predictable resources. With tight budgets, the freedom to experiment and to improve without incurring unbearable costs enabled by a flexible genome remains an illusion. Resources allow for the full expression of the versatility inherent in the most adaptable genomes, languages, and industrial protocols. A growing wealth of food, money, genes, words, and techniques enables—but does not compel—the user to improve on many fronts at the same time. Implementing new technology, especially the kind that requires large infusions of energy, is precluded under scarcity no matter how sophisticated the genome or the benefits of a fully realized adaptive state.

Although the inanimate Earth is the ultimate source of the mix of substances that living bodies incorporate, organisms deliver and control these essential constituents. Without photosynthesis, there would be no free oxygen in the atmosphere or ocean. Without bacteria and fungi that dissolve and break down minerals in rocks by chemical weathering, there would be no soil on land, and plants would be unable to extract needed nutrients from it. Rivers would empty into the sea, but their waters would be low in runoff to fertilize marine life. Without animals stirring up sand and mud on the seafloor or turning over the soil on land, valuable nutrients would be rapidly and permanently buried, so that any new minerals brought up by volcanoes or in hydrothermal vents would soon be lost. If there were no land plants transpiring water into the air, many places on Earth would receive less rainfall and be less productive than they are. Not only does life capture the raw materials that physical forces in Earth's mantle and crust make available, but it recycles and redistributes them. By modifying and intervening in Earth's surface chemistry, it cumulatively establishes an environment in which the building blocks of life become increasingly plentiful, metabolically accessible, and predictable. In other words, life constructs and regulates its own resource economy.

Resources and an evolved organization favorable to innovation provide freedom and opportunity by loosening constraints on activity, but they promise only potential and possibility. To transform option into reality, there must be selection, the sorting of living things according to performance criteria set not only by the inanimate environment, but even more by life itself. Greatest access to locally scarce resources, and greatest physiological control of the body's internal state, are afforded to those organisms that, individually or collectively, do the most work in the shortest period of time. The advantages of power are many. A competitor can acquire and defend more resources, gain independence from its variable surroundings, and gain access to sites of plenty as well as to locations of safety from enemies. Compared to less powerful agents, it exerts more control over the activities, distribution, resources, and adaptive characteristics of the life-forms with which it interacts.

There will thus always be selection in favor of greater power and influence when individuals compete for scarce resources. To the extent that constraints on power are lifted through increases in versatility, or the range of adaptive possibilities, this relentless selection could result in an overall trend toward more powerful competitors among the ranks of the winners, organisms or groups that exercise disproportionate control over others.

The physical dimensions of power indicate which characteristics confer a competitive edge. Power—energy acquired or expended per unit time, measured in watts—is enhanced when mass, distance of action, exerted force, speed, acceleration, or available energy increase, or when the time interval during which a given task is accomplished decreases. Power is expensive; its benefits accrue only to those who are able to spend profusely. Well-trodden pathways to power have led members of many lineages to become large, fierce, venomous, fast-growing, keenly sensitive over large distances, quick, productive, cooperative, long-lived, flexible, internally highly regulated, warm-blooded, metabolically active, or intelligent. Every avenue toward increasing power has been traveled by multiple lineages no matter from which point these lineages depart. Regardless of the aspect of power one chooses, lineages living today rank at or near the top of every category; they exceed the capacities of most of their more ancient counterparts.

When we plot these various expressions of great power on the evolutionary tree, an unmistakable pattern emerges, which implies that the criteria for competitive superiority have remained unchanged even as the performance standards have increased over the course of the biosphere's history. With the possible exception of weedy opportunists, which project power mainly through brief rapid growth and high fecundity during a short life span, the characteristics conferring power and permanence are concentrated near the tips of the tree's branches. Organisms representing the ancestral condition of low power are found throughout the tree. Thus large creatures came from small ones, but many lineages remained small-bodied. Color vision came from simpler light perception, and image-forming eyes came from eyespots; but unsophisticated visual systems and the absence of eyes are spread throughout the animal part of the tree of life. Closed hydraulic systems—the network venation of flowering-plant and many fern leaves, the veins and arteries and capillaries of our circulatory system—evolved from openly branched ones, but the unchanged ancestral condition persists on the tips of many branches on the tree of life. Plants whose basal growth enables them to tolerate intense continuous herbivory came from ancestors whose apical growth made them vulnerable to such damage; but apical growth persists in many lineages. Predators using force or venom to bring down large or obstreperous prey are derived from smaller, more passive consumers that took much smaller and less aggressive victims, but again the less powerful methods of predation work well for vast numbers of living animals. Insect societies acting as highly integrated groups with castes specialized for different tasks evolved from less differentiated societies of fewer individuals, and these in turn arose in lineages of solitary insects; but many species are successful without being social. Intelligence, with all the cognitive intentionality it entails, is evolutionarily rooted in states in which awareness was both more local and more limited to the present.

It would be wrong to conclude that these states of power are irreversible. In fact, power is far more often lost than gained, because the number of losers far exceeds the number of winners in competitive contests. Careful studies of the tree of life reveal innumerable cases of simplification and few transitions to greater complexity. Size can decrease as easily and as frequently as it increases. Hundreds of lineages in caves and the deep sea have lost eyes and have therefore become blind. Countless insect and bird lineages have become flightless, whereas the capacity of powered flight evolved not more than four or five times. And although there appears to be no example of reversal once a differentiated caste structure has evolved in insects, earlier stages in the evolution of social insects can be reversed, as indicated by the position of some social bees at the base of lineages of solitary species.

Many evolutionary biologists would consider these cases of reversal as proof that there is no tendency for powerful agents to be concentrated in outer parts of the tree or for any claim of a general upward trajectory of power. Reversals, however, imply a fall from power. In an economic as well as an evolutionary sense, a gain in power has a greater effect on the selective regime, distribution, and activity of organisms than does a decrease. Individual organisms are not equivalent to each other in the economy of nature; neither are evolutionary gains and losses of structures.

An even graver error would be to say that all tips of branches in the tree of life are occupied by powerful species. Every living lineage represents a tip, and only a few lineages comprise powerful elites in their communities. A more accurate depiction of the facts is therefore that states of low power are spread throughout the tree, near the bases of most branches as well as on many tips, whereas powerful agents are concentrated near the ends of branches. Moreover, some low-energy occupations represent evolutionary cul-de-sacs, the pathways to which are strictly one-way. This is because the competitive dominants have pushed vast numbers of lineages into ecological refuges whose asylum status—greater safety, fewer and less accessible resources—has been stable relative to the source environments in which evolutionarily more volatile species with power operate. The tree of life shows that lineages evolve from seafloor dwellers to rock-excavating or planktonic creatures, from free-living organisms to parasites or mutualistic symbionts, from warm-adapted species to polar ones, from shallow-water to deep-sea animals, and from continental to island life. Few if any cases of evolution in the opposite ecological direction are known. A few deep-sea corals, clams, and crustaceans have surfaced in shallow waters, but only in cavelike or polar habitats where the resource regime and predatory environments resemble those in the abyss.

If power and resources were independent of each other, there would likely be no historical pattern other than an endless succession of cycles. Selection would ensure the evolution of powerful agents, but disruptions remove those with the greatest power from time to time, potentially wiping out accumulated gains and allowing the process to begin again as surviving lineages take the place of those that succumbed. A fixed upper limit on power might thus remain in force for vast stretches of geological time. In fact, organisms might even deplete their resources, so that the potential for attaining great power actually declines over successive generations. For example, trees in forests deplete soils of soluble minerals, and are gradually replaced in successive generations by trees of lower stature unless soils are replenished with nutrients in the form of dust carried on the wind or mud deposited by flooding rivers. Such depletion characterizes forests on the Hawaiian Islands, which receive little outside dust, as well as in cold or dry places like Alaska, northern Sweden, and Australia.

But power and resources are often linked through strong positive feedback. Compared to organisms that deplete their resources, life-forms that stimulate the production of the commodities they need gain a competitive edge by growing faster and by lifting other limitations. Reinforcement between living things and their resources allows power to accumulate and the scope of adaptation to expand. The counterintuitive consequence is that the most effective competitors, rather than destroying or diminishing the economies in which they operate, strengthen them through interdependencies, which push back the walls and raise the ceiling of constraint. They conspire with their victims to establish an economy that moves life, slowly and haltingly, toward greater productivity, power, and internal control.

Evolutionary escalation illustrates these feedbacks well. Recall from chapter 6 the evolved relationships among predators, shell builders, and secondary shell occupants. Predators select for a passive shell defense in their prey, but also make intact shells available to secondary shell dwellers. The resulting increase in the resource of shell-encased prey creates an even more lucrative target for predators specializing on that resource. In chapter 7, I suggested that herbivores often increase the productivity of their plant food by fertilizing plants and by speeding up the time it takes for plant tissue to pass from plant to consumer and back again. This positive feedback between consumer and producer sets the stage for the evolution of herbivores with bigger appetites. The symbiosis between root fungi and land plants doubtless provided a competitive edge relative to plants without roots or without fungi. The recent spread of grasses, facilitated by hungry mammalian herbivores, likely contributed substantially to this increase in chemical weathering and productivity. The silica extracted from the soil by grasses for structural defense is exported in dissolved form by rivers to the ocean. There, it has fueled a rise in marine productivity by silica-using diatoms.

Chemical weathering also releases calcium. In the ocean, nearly all the dissolved calcium ends up in the calcium-carbonate skeletons of animals, seaweeds, and planktonic organisms. These skeletons in turn comprise most of the world's limestone. Beginning about 550 million years ago, just before the dawn of the Cambrian period, at least twenty-one lineages of marine organisms evolved the ability to secrete mineralized skeletons. One of the earliest organisms with a skeleton, the tubular fossil *Cloudina,* already shows indications of unsuccessful predation in the form of small holes that were drilled from the outside wall but sometimes did not penetrate to the tube's interior. This observation and other fossil occurrences in the Early Cambrian period indicate that skeletons served as antipredatory devices from the very beginning. Early lineages with skeletons were found mainly on the shallow seafloor, but they gradually spread to deeper waters and later to the open ocean. The first planktonic organisms with a skeleton of calcium carbonate may have appeared as early as the Silurian period, about 420 million years ago, but open-ocean plankton with such skeletons became abundant much later, beginning in the Early Cretaceous, about 120 million years ago, but especially from about one hundred million years ago onward. Regulation of calcium by organisms thus extended from shallow waters to the entire ocean. In corals and some mineralizing seaweeds, skeletal formation stimulates photosynthesis. The domestication of the calcium cycle by life on Earth is thus due both to predators as selective agents and to some very important primary producers.

The production of oxygen through photosynthesis was at first a disaster for the anaerobic microbes populating the biosphere, because free oxygen is toxic to these organisms; but it eventually opened up opportunities and permitted a vast increase in both power and the availability of resources. Aerobic metabolism—essentially the reverse of photosynthesis, and therefore a process that consumes oxygen—yields up to forty times as much energy-storing ATP (adenosine triphosphate) as do forms of metabolism that operate in the absence of oxygen. This greater access to energy became possible when oxygen levels reached about one-hundredth of the modern level of 21 percent by volume of air some 2.4 billion years ago. It was not until about 580 million years ago that oxygen became plentiful enough in the atmosphere and ocean to sustain large organisms up to 1.2 meters (about four feet) in length. Some of these larger life-forms may have been among the first animals—a major episode of diversification of animals began with another rise in oxygen levels 550 million years ago, just before the Cambrian period. For the first time, living things began to burrow into the microbial mats that had covered the seafloor until the Cambrian Period. Nutrients that had become entombed within and beneath these mats could now be recycled thanks to the activities of burrowing animals. With more resources made available to the ecosystem, productivity must have risen. Oxygen levels rose yet again with the evolution of land plants, making possible the evolution of powered flight, first in insects, about 320 million years ago, and later—during the Late Triassic, about 225 million years ago—in pterosaurs. Birds independently took to the air seventy-five million years later. Warm-blooded physiology, requiring copious oxygen, evolved in several vertebrate lineages during the Late Triassic period. Progressive oxygenation of Earth's surface environments eventually opened up the deep sea to aerobically metabolizing animals, and therefore permitted ever more recycling of resources.

Meanwhile, the evolution of swimming, walking, and flight promoted nutrient exchange, carried on by animals, and the spread of rigorous selective regimes. By roaming widely in search of prey or other sparsely scattered food, animals like whales, tunas, pollinating bees, and migrating birds—all evolving in the past one hundred million years and, in the case of the most wide-ranging and fastest species, in the last thirty million years—cross boundaries between regions and thus effectively connected natural economies that would otherwise have operated as separate, smaller units. Through their wanderings, these animals globalized selective regimes and established trade routes of nutrient exchange and subsidy among distant ecosystems. As with human trade, natural economies and the species in them benefit from these patterns of subsidy, becoming more productive than they would have been in the absence of these vagrants.

The evolution of predators, an event of immense importance that ushered in the Phanerozoic eon and thus the consumer age some 542 million years ago was another consequence of the greater availability of oxygen made possible by photosynthesis. The earliest predatory activities, recorded as complete and incomplete holes in the tubular skeleton known as *Cloudina,* date from 548 million years ago, just before the beginning of the Cambrian period. Walking, crawling, and swimming predators were already diverse in the Early Cambrian, including claw-bearing basal arthropods known as anomalocarids, which reached a length of a meter or more. Successive top predators in the sea became increasingly powerful. The Late Devonian placoderm fish *Dunkleosteus terrelli,* living some 370 million years ago, was the first predator capable of capturing large evasive prey and killing it with a strong bite. At a length of six meters, this fish is estimated to have exerted a bite force of 5,300 newtons with tooth plates at the rear of the jaws. This force compares to 550 newtons for a large dog and 749 newtons for a human bite. The most powerful predator that evolved before humans dates from the last twenty million years. The enormous shark *Carcharodon megalodon,* with sharp-edged teeth twelve centimeters long and an estimated body mass of 103,000 kilograms, had a calculated bite force of up to 182 thousand newtons, powerful enough for a diet of large whales. By comparison the nearest living relative, the great white shark *(Carcharodon carcharias),* reaches a maximum body mass of 3,324 kilograms and bites with a force exceeding eighteen thousand newtons. The killer whale may be able to muster a bite force comparable to that of *C. megalodon,* but its bites have never been measured or estimated. This current top marine predator evolved no earlier than the Pliocene epoch, at most five million years ago.

This dramatic increase in strength among top marine predators is paralleled by other consumers. In chapter 3 I noted that predators using force to break or enter shells became increasingly powerful over time, culminating with such animals as pufferfish, large rays, and the loggerhead turtle. As I discussed in chapter 7, there is a similar case to be made for successive top herbivores, both on land and in the sea. The record of land-dwelling top predators appears to show a comparable trend. The largest known species during the Late Permian, Triassic, Late Jurassic, and Cretaceous periods weighed approximately 400,600, more than 1,000, and 5,700 kilograms respectively. *Tyrannosaurus rex,* the largest Late Cretaceous predator, is estimated to have had a bite force of 13,400 newtons, higher than that of the most powerful living mammal so far measured (the African lion, 4,168 newtons) but comparable to the American alligator (13,300 newtons). Post-Cretaceous predatory mammals are all much smaller than their earlier dinosaur counterparts. The largest living carnivore is the Asian tiger (272 kilograms), but Pleistocene tigers in Java may have been half again as heavy. Cats living during the Late Miocene to Pleistocene reached a body mass of 400 kilograms or more, and the

Late Eocene predatory mammal *Sarkastodon* from Mongolia is estimated to have weighed 800 kilograms. Although no predator of the last five million years comes close to this mass, it is notable that many of the top predators of this most recent interval in Earth's history are group hunters, which can bring down larger prey than more powerful individuals of solitary species can.

In short, consumers and resources have become increasingly intertwined through time. Photosynthesis led to oxygen, which permitted mobility, which led to nutrient recycling and to larger, more productive ecosystems; and these systems supported more powerful consumers, which increased the rate of supply of resources still more.

The great biosphere-scale crises—that is, the mass extinctions—neither interrupted these trends nor disrupted the feedbacks that made them possible for long. Top consumers were particularly prone to extinction, with the result that established interdependencies came undone and intense selection due to consumers relaxed temporarily; but their places were taken by other, less powerful predators, which quickly assumed the role of top dog and often surpassed their predecessors in power. At each of the mass-extinction events, there was a temporary halt in the production of marine limestone, indicating a severe cutback in the abundance of calcifying organisms and a dramatic jolt to the calcium and carbon cycles; but the limestone gaps lasted for only a short time, and the general trend for fossil beds of shells to thicken (and thus to become more productive) over successive geological intervals was scarcely interrupted. The players changed, but the historical patterns did not.

This view of history implies neither a straight pathway of "progress" nor a denial of contingency. Directionality emerges as a general, cumulative trend, initiated and propelled by competition and diffusing throughout the network of life and its resources. It is driven not only by evolution within lineages, but also by the eventual replacement of extinct elites with even more powerful members of previously subordinate lineages. We cannot know in advance who or what acts how, where, when, or why, nor can we predict or even describe the particular situations and events in the ever-changing network. These remain in the realm of unpredictability and contingency. What we can know, and predict, are adaptive characteristics that are favored during episodes of selection in which the acquisition and defense of resources set the criteria of performance, because these properties benefit their bearers in frequently occurring circumstances. These characteristics—a high rate of metabolism, social organization, intelligence, high productivity, and so on—indicate intense economic activity and enable their bearers to exercise strong economic and evolutionary control over other life-forms. Life through adaptation creates its own predictability and expands its own opportunities. In the long run, it shapes its arrow of time, not through intentional action or a guiding hand, but by universal selection. The details of time, place, and player reside fully in the realm of inscrutable chance.

This is all well and good for nonhuman life, but how do humans fit into this picture of increasing power and productivity? And what will our future hold? What lessons can evolutionary history teach us about ourselves? These are the questions to which I turn in the final chapter.

CHAPTER THIRTEEN

History and the Human Future

The Netherlands is one of the most densely populated countries in the world, but as a boy growing up there in the 1950s, I was rarely far from nature. The pines and beech trees on the grounds of the institute for the blind in Huizen offered quiet relief from the noise and oppressive discipline of the house in which I lived with eleven other children. On those rare weekend occasions when I was home in Gouda, we often walked or bicycled to the surroundings polders, where the lush meadows were full of flowers and the muddy ditches were choked with interesting water plants. Gouda, the only town I knew well, was a compact, convivial place that, despite its narrow streets and crowded cobblestoned market square rarely gave me the feeling of being boxed in by the human throng.

I was hardly prepared for a city like New York. Our first visit there was in the fall of 1955, a few weeks after we arrived in America. To say that I was overwhelmed by the traffic, the din, the pollution, and the sheer size of Manhattan would be an understatement. Nature had not been tamed here; it had been obliterated. Everything here was human on a grand scale, from the subway tunnels underground to the tops of the skyscrapers. New York seemed like the center of the world, a place of promise and economic opportunity where millions came to make fortunes and a better life. The city was a monument to economic power, a brash symbol of human domination over the Earth.

In the middle of a big city like New York, Lima, or London, it is difficult to remember that we humans are part of nature, that we are the product of the same evolutionary processes that gave rise to all other forms of life on Earth, from bacteria and birds to corals, clams, and clovers. And yet we are. Our ancestry is deeply rooted among the ranks of animals, and our evolution represents a continuation of historical trends toward greater power among winners in the perpetual competitive race that has gone on since life's beginnings on our planet.

We are connected to the past, but how good a predictor is history? How much of the future is really in our hands? Can we continue to adapt? With perhaps more hope than conviction, I suggest that we shall continue along the trajectory laid out when life began, a trajectory set in motion by selection in favor of living things whose emergent goals are to survive and propagate. We, like the rest of the living world, are subject to the forces of competition and cooperation that dictate how locally scarce resources are divided among organisms and among parts of organisms. Unless we replace survival and propagation with other criteria for success—a profoundly unlikely prospect—we cannot escape this reality, and we shall remain bound to history.

This perspective on history as prediction, however, deserves close scrutiny. Many social policies and aspirations would, if they came to pass, represent real departures from historical precedent. The possibility of perpetual growth envisioned by economists, the replacement of competition by cooperation imagined by some utopians, and a hoped-for trend toward a reduction in the use of natural resources by us, the dominant species of our age, all conflict with long-established trends in the history of life. A decline in species diversity in the human-dominated biosphere would also apparently reverse a well-established pattern, in this case a general increase in the diversity of multicellular life. History would, in other words, be a poor guide to the future if any of these future states were realized.

The question before us is simple yet profound: is the predictiveness of history so powerful that the human future as conceived by economists and utopians is unachievable in principle, or can collective decisions release humanity from the yoke of the past? No matter what the answer is, we have the responsibility to place the past and the laws of evolutionary change at the center of how we craft the future.

It is easy for a lover of nature like me to see the destructive side of human activity. In the wake of our relentless economic expansion, the diversity of life-forms on Earth is declining. Many ecosystems are becoming islandlike refuges, and others are turning into globalized mixtures of human-tolerant species. Earth's topography is being profoundly altered: coal companies are leveling mountains and filling in valleys; man-made canals link previously unconnected seas and river systems; and breakwaters change coastal patterns of sand deposition. Most big rivers and many smaller streams are dammed; forests and grasslands are being transformed into agricultural and urban landscapes. Trawling and dredging are decimating rich bottom faunas on the world's continental shelves. Our burning of fossil fuels is warming Earth's climate to a state not seen since the Pliocene epoch, three million years ago. And the list goes on.

Putting aside these destructive aspects for the moment and looking at our place in history from a human-centered and species-selfish position, I suggest that *Homo sapiens* falls in line with earlier top consumers and producers in propelling ourselves and the ecosystems we control toward greater productivity. With the invention of synthetic fertilizers and the burning of fossil fuels, the supply of fixed nitrogen—the nitrogen that organisms can use—has more than doubled globally. Figures quoted by the University of Minnesota's Ford Denison and his colleagues indicate that a productive almond orchard yields up to 1,800 kilograms of nuts per hectare, more than 230 times higher than the maximum 55 kilograms of acorns produced in an oak forest. Over a three-year period, the average agricultural grain yield (almost 2,500 kilograms per hectare) is five times higher than the mass of seed in the most productive prairie. Productivity in coastal zones of the ocean worldwide has doubled thanks to soil erosion and farm chemicals rich in nitrogen and phosphorus flowing into rivers and the sea. The release of nutrients through fires, the frequency of which increased beginning seven to eight thousand years ago as human agriculture was spreading, has heightened plant productivity on land. So has the increasing concentration of carbon dioxide in the atmosphere, the current level (about 370 parts per thousand) of which is higher than at any time since at least the Early Pliocene, three to five million years ago. Although irrigation eventually makes soils salty and therefore unfit for most agriculture, it has transformed relatively unproductive natural ecosystems into fertile ones. Some 40 percent of our food comes from irrigated farmland, which comprises just 16 percent of total land area. The per-capita production of food has outpaced even the 10-fold increase in the human population since the year 1600 and the 2.5-fold increase since 1955. Although the standing biomass on farmed and grazed land is much lower than that on the grasslands and forests that these managed ecosystems replaced, productivity there is markedly higher.

Over the past five hundred years, the human population and its economic stamp on Earth have grown so fast and so consistently that most economists, and likely also the great majority of the public, have come to expect that this growth will continue indefinitely. Growth, in this view, is not just the natural state of things, but the only desirable one. The thinking that undergirds this belief seems to be that, as one critical resource is depleted, another equivalent one will take its place. Pollution and climatic consequences aside, a plausible argument for such substitution can be made for fuels: wood from forests was replaced by coal, which in the late nineteenth century was complemented with petroleum, which is now being partially replaced by natural gas, coal, living biomass, and nuclear materials. But how good is the prospect for substitution of such other resources as food, potable water, and living space? Will we bump up against an economic plateau, or will ingenuity and technology allow us to break out of our resource constraint?

Many economists deny the premise of these questions. To them, the human economy is no longer limited to tangible resources like food, water, land, and fuel. The following comment from a respected economic historian is typical: "The force constraining the expansion and growth of human society is not the supply of resources, natural or otherwise, but the exhaustion of its dynamic strategies and the inability to replace them with new strategies."[1] My former colleague at the University of Maryland, the late economist Julian Simon, argued that supplies of natural resources will increase indefinitely because as the human population expands, the chance that some brilliant genius will solve the problem of diminishing supply will be close to 100 percent. Knowledge and money are indeed necessary to discover, unlock, and make available resources and to use them effectively; but they must be applied to something tangible. We are, and presumably will remain, made of organic matter, and we live because we metabolize; and all of this depends on the ready availability of life's essentials. To deny this, it seems to me, is to engage in economic alchemy.

Although there has been an overall rise in both consumption and production over the course of the history of life on Earth, the increase has not been uniform, but instead seems to have been concentrated during a few notable periods. Most prominent among these are the times around the beginning of the Cambrian period when muscle-powered animals including predators first appeared, and the mid- to Late Cretaceous, when highly productive flowering plants expanded together with social insects and warm-blooded, or at least very active, vertebrates. Other intervals of worldwide economic growth likely include the Late Silurian to Early Carboniferous, when land plants and marine predators greatly expanded, and the Late Triassic, when a great number of adaptive innovations associated with higher power arose, including vertebrate flight (in pterosaurs). Even in the last 65 million years, some intervals of global warming—the Late Paleocene to Early Eocene (56 to 52 million years ago) and the Late Oligocene to Middle Miocene (27 to 16 million years ago) may have witnessed overall growth.

The unusual conditions that make major global growth possible are, first, an external trigger—intense volcanic activity or an episode of mountain-building resulting in complex, erosion-prone terrain—that releases new essential resources from deep within Earth's crust and mantle; and, second, the existence of organisms that are capable of capitalizing on those resources by forming cooperative partnerships and creating positive reinforcement between consumers and producers. Over the course of geological history, organisms have "learned" to mine previously inaccessible nutrients by extracting them from soils and sediments, capturing unusable nitrogen from air to make nitrates and other compounds to make proteins, and releasing essential minerals from rocks. But humans have taken these measures to new heights. Using fuels mined from deep within the Earth, we are manufacturing fertilizers rich in nitrogen and phosphorus. In effect, we have taken control of processes of resource supply that previously were dominated by geological forces. The perpetual growth that economists and the public imagine for the future of our species would thus be possible only if such unprecedented resource extraction, together with our ability to capitalize on the new supply, continued unabated.

But is this realistic? Thoughtful analyses by the University of Minnesota's Ford Denison and his colleagues indicate that the rate of improvement in yields of such important crops as rice, wheat, and corn has been slowing over the last thirty years despite great biotechnological advances. According to these scientists, future increases of more than 20 percent over present yields are unlikely without some unforeseen breakthroughs in forging much more efficient enzymes involved in photosynthesis. It therefore seems to me that, unless we expand our reach beyond the Earth, our species will sooner or later reach a plateau beyond which growth will be at best sporadic. Even to get to that point, we would have to exercise more control over the Earth and levy still higher taxes on ecosystems than we are already doing. I am too much in love with nature to find such a prospect appealing.

Before humans evolved, the chief architects and consumers in ecosystems created habitats such as the forest understory, the safe bodies of dangerous predators, and crevices among corals, where a host of species that could not have existed without this infrastructure make a living. Subject to the limitations of climate, top competitors thus created opportunities for species and promoted diversity at the level of species. In other words, diversity begets more diversity, mainly through the evolution of highly habitat-specialized parasites, mutualistic partnerships, and organisms living low-energy lives.

Humans have broken with this tradition. Our activities are so invasive that diversity, both regionally and globally, is declining, especially among animals and plants of large size. And yet diversity thrives in our midst. It is not diversity at the species level, but the diversity of occupations within our own species that has risen so dramatically during our history. The invention of agriculture, and later of urban life and industrial civilization, led to the creation of a multitude of new occupations that could not have been imagined in earlier phases of our cultural development. We have created a social infrastructure in which people can make a living as soldiers, stockbrokers, bishops, auto mechanics, teachers, composers, nurses, shopkeepers, housemaids, police detectives, computer technicians, welders, and consultants. Human occupations are, in economic terms, comparable to species in nature. In effect, the locus of diversity has begun a shift from species in ecosystems to occupations within the confines of a single overdominant species.

This transformation certainly represents a profound break with a long-established historical precedent, but it is not the first such transformation. The species, whether we think of it as a genetically circumscribed group of individuals isolated from all others as a twig in the evolutionary tree of life or as an ecological unit with a particular way of life dictated by enemies and resources, is an entity that emerged only when sexual reproduction and the vertical transmission of genes from one generation to the next became the norm. Before the emergence of species, cells widely shared genes through horizontal transmission, and evolution occurred largely at the community level rather than within the confines of lineages. Nearly all important organic compounds had evolved during this early microbial phase, before sexual reproduction became established. Only lignins—polymers that stiffen land plants—and some defensive substances of plants and venoms of animals were added to the array of organic compounds synthesized by early life-forms. Much of the molecular variety created during the early phases of life's history persists in the microbial realm, but larger life-forms employ only a fraction of it. Instead, these larger organisms—especially plants, fungi, and animals—built an edifice of diversity at the species level, a diversity emphasizing form rather than chemistry. The locus at which diversity is expressed thus shifted from the chemical to the morphological with the advent and rise of multicellular life. Now humans are again changing the locus of diversity, from that at the level of species to the level of occupations and products within our own species. In other words, diversity increases under each successive evolutionary regime, from the microbial to the multicellular to the human, but the level at which it is expressed has changed. If we can promote diversity within our own species without destroying the capacity of other life-forms to maintain and generate diversity at the molecular and species levels, we shall follow the great arc of history by continuing the trend toward greater variety and creativity that has characterized the history of life.

A far more fundamental departure from history would occur if destructive competition could be replaced by cooperation and harmony. Competition diverts resources from useful ends to weapons, and in its traditional form promotes selfish ends over communal ones. There are many who either wish competition away or deny its universality. Following a long historical tradition, the British economic historian Geoffrey Hodgson maintains that "[n]either biology nor anthropology give support to the universal presupposition of competition and scarcity."[2] I disagree with both Hodgson and with the assertion that cooperation eliminates competition. It is true that a critical resource like oxygen is in no sense limiting the human population or indeed the biosphere as a whole, but like other resources, it is often locally scarce, as in the sediments of salt marshes or in unventilated burrows in the ground. Competition is local. What matters in evolution and in the human economy is whether individuals or groups can acquire and defend commodities as they interact with other life-forms. Selection due to competition is local, powerful, and universal.

What would the world look like if there were only cooperation and no competition? It would, I think, be a very precarious place, one that would not last very long. For humans, all our decisions and actions would be agreed upon unanimously, or imposed by a single dictator; but there would be no guarantee that they would be the right ones. The absence of dissent is equivalent to having a single hypothesis, a monopoly, without alternatives that can be tested against one another. Monopolies work as long as they have the intended good effect, but they fail catastrophically if they make a mistake, as seems inevitable in the long run given that time and knowledge are never sufficient to consider all the possible consequences.

In the economy of life, cooperation and competition are inextricably linked. Cooperation within a group, or between two species, creates an entity that is itself a better competitor than the parties would be if they acted alone. As a highly effective form of competition, cooperation pushes conflict to a higher level of organization. Competition in the absence of any cooperation is destructive and uncreative, and leads to retrenchment and conservatism; but competition in the form of pitting alternatives against each other is an essential, emergent process coupled with selection and adaptation. We should strive toward making competition within society and between humanity and the rest of the biosphere less violent, but we cannot and should not eliminate it. If we did dispense with competition in our daily lives, we would squelch all individual incentive and ultimately court monopolistic Armageddon.

Through much of human history, wealth and power have been equated with material accumulation. The rich build palaces, adorn themselves with expensive jewels, drive fancy cars, entertain lavishly, travel far and wide in style, and fill their closets with too many clothes and shoes. Priceless works of art decorate their houses. The link between status and materialism, or at least conspicuousness, arises from competition and has deep evolutionary roots. Flowers must be eye-catching or fragrant to be noticed by pollinators; and male Anna's hummingbirds expend vast amounts of energy as acrobatic aerial divers to impress potential mates with their health and vigor. If we could sever the link between prestige and material plenty, we humans could temper our collective use of resources without significantly compromising our standard of living.

And here I see a glimmer of hope. Beginning with Andrew Carnegie, a ruthless industrialist who exploited his workers and accumulated a vast fortune in steel and railroads, but who late in life gave away much of his wealth, many of the superrich have become conspicuous philanthropists, putting their wealth and influence to use in building libraries (one of Carnegie's legacies), endowing universities and hospitals, lifting the poor out of poverty, funding scientific research, supporting museums, and giving scholarships to needy students. Directly and indirectly, I have been the beneficiary of many of these efforts. The emergence of philanthropy represents a recent but important shift in values, in which prestige is increasingly measured as contributing to the common good rather than as just luxurious living.

This does not mean that the rich and powerful will yield their wealth and influence to others voluntarily. Forced redistribution of wealth through expropriation of land, taxation, and wage reduction is always strongly resisted unless those affected perceive a clear benefit. Losses are usually weighed more heavily than gains. Only if the total resource base expands, as happens when an economy grows, will people tolerate some redistribution, for under those conditions individuals will not lose even as others gain. These realities of human psychology have deep roots: no organism will voluntarily cede power or biomass without some tangible threat. Lizards and salamanders may throw off their tails when confronted by a predator, and aggressive crabs may leave a claw pinching my thumb as they escape, but their evolved sacrifices are preferable to death.

Taking this argument to the level of the human species as a whole, I doubt that our collective use of energy will decrease without some kind of coercion, either by physical forces beyond our control or from internal conflicts. Our only hope is a change in values, an adaptive modification of the social hypothesis that makes our species work.

Human values are neither static in time nor uniform across societies. Most human groups no longer accept slavery, child labor, capital punishment, and torture, practices that are common to nearly all societies in the past. Appalling transgressions occur on all sides, of course, but there has been a general shift away from these abuses. Even religions, the bastions of social conservatism, have evolved over the centuries, splintering and diverging like species and occupations. As hypotheses of the human social environment, they reflect prevailing social norms, but like living adapted organisms, they also affect those norms.

As I write this, a concerto grosso by Georg Friedrich Händel is playing on the radio. Elegant objects of natural history—shells, pinecones, and twisted seed pods—each with their own stories of adaptation and evolution, adorn my room. Mockingbirds and white-crowned sparrows sing outside my window, and the fragrance of climbing *Clematis* blossoms perfumes the spring air in our garden. None of this beauty was created to please me, but I perceive it as uplifting and enriching nonetheless. It is that aesthetic dimension of love, that emergent consequence of basic competition among living things for locally scarce resources to carry life forward, which is for me one of the two primary values that infuse life with transcendent meaning. The other is scientific understanding and the potential good it can do. There is power in the idea that the struggle for life can lead to so much variety and creativity; and there is unity in the phenomenon of adaptation, which through emergence and selection is the foundation of evolution and of the scientific way of knowing. We are part of the edifice of life that is built on that foundation, and it is our duty to see to it that our house—our *oikos,* our economy and that of the other occupants of the biosphere—is kept in order and fit to live in.

Each of us has but one life to live on one habitable planet. It is both a privilege and a responsibility to celebrate, understand, and protect this world.

SUGGESTED FURTHER READING

Chapter 1: The Evolutionary Way of Knowing

A detailed discussion of adaptation is given in chapter 3 of G. J. Vermeij, *Nature: An Economic History* (Princeton, NJ: Princeton University Press, 2004). In that book, I make extensive use of the idea of adaptation as hypothesis. This idea seems to have been proposed first by Wolfgang Sterrer in "Prometheus and Proteus: The unpredictable individual in evolution," *Evolution and Cognition* 1: 101–129 (1992).

Human bipedalism is beautifully described and explained by D. M. Bramble and D. E. Lieberman in "Endurance running and the evolution of *Homo*," *Nature* 432: 345–352 (2004).

Cases where traits reflect either construction or a real disadvantage to their bearers have been widely discussed. See, for example, the classic paper by Stephen Jay Gould and Richard C. Lewontin, "The spandrels of San Marco and the Panglossian paradigm: A critique of the adaptationist programme," *Proceedings of the Royal Society of London B* 205: 581–598 (1979); and George C. Williams's book *Natural Selection: Domains, Levels, and Challenges* (New York: Oxford University Press, 1992). The best source on neutral evolution is still

M. Kimura, *The Neutral Theory of Molecular Evolution* (Cambridge, UK: Cambridge University Press, 1983).

The distinction between exploration and selection has received thorough attention in two excellent books: M. Kirschner and J. C. Gerhart, *The Plausibility of Life: Resolving Darwin's Dilemma* (New Haven: Yale University Press, 2005); and J. S. Turner, *The Tinkerer's Accomplice: How Design Emerges from Life Itself* (Cambridge, MA: Harvard University Press, 2007).

Although I disagree with his discussion of adaptation, Peter Corning has written a deeply important book on evolution, with particular emphasis on the concepts of emergence, synergy, and the purposiveness of survival and propagation. Anyone interested in evolution, economics, and the human species should read his *Holistic Darwinism: Synergy, Cybernetics, and the Bioeconomics of Evolution* (Chicago: University of Chicago Press, 2005).

David L. Hull's *Science and Selection: Essays on Biological Evolution and the Philosophy of Science* (Cambridge, UK: Cambridge University Press, 2001) is an excellent and thoughtful introduction to the study of the scientific enterprise from an evolutionary point of view. Hull does not, however, draw the parallel between science and adaptation. His most important contribution is the idea that science usually acts in the best interests of society as a whole, because scientists' self-interests favor honesty, accuracy, and openness.

Stephen J. Gould has written extensively about the illusion of progress. See his book *Wonderful Life: The Burgess Shale and the Nature of History* (New York: W. W. Norton, 1989), which is as eloquent a source as any.

The idea of religion and other doctrines of allegiance as social adaptations is most thoughtfully discussed by David Sloan Wilson in his *Darwin's Cathedral: Evolution, Religion, and the Nature of Society* (Chicago: University of Chicago Press, 2002). The claim that religion is little more than a by-product of the ability to read other people's minds seems to me at most an incomplete explanation. Particularly strong proponents of this view are Daniel C. Dennett, *Breaking the Spell: Religion as a Natural Phenomenon* (New York: Penguin, 2006); and Richard Dawkins, *The God Delusion* (New York: Houghton Mifflin, 2006). Arguments presented by these two authors on the flawed notion of intelligent intervention in evolution are sound. The most compelling case against intelligent design is Michael Shermer's *Why Darwin Matters: The Case Against Intelligent Design* (New York: Henry Holt, 2006). Stephen Jay Gould, in *Rocks of Ages: Science and Religion in the Fullness of Life* (New York: Random House, 1999), recommends a strict separation between science and religion; but by my reading he rightly seems to relegate the religious "way of knowing" to a tiny principality rather than to elevate it to Gould's imagined "magisterium." An informative little essay on the origins of European science is Philip Ball's "Triumph of the medieval mind," *Nature* 452: 816–818 (2008).

The narrative of the northern oceans is summarized in G. J. Vermeij, "Community assembly in the sea: Geologic history of the living shore biota," pp. 39–60 in M. D. Bertness, S. D. Gaines, and M. E. Hay (eds.), *Marine Community Ecology* (Sunderland, MA: Sinauer Associates, 2001).

Chapter 2: Deciphering Nature's Codebook

The requirements that any living system must satisfy in order to sense and interpret its environment have been carefully spelled out in Rupert Riedl's *Order in Living Organisms* (translated by R. P. S. Jeffries) (New York: John Wiley and Sons, 1978). A highly readable summary of the evolution, elaboration, and convergence of sensory capacities in animals is given by Simon Conway Morris in *Life's Solution: Inevitable Humans in a Lonely Universe* (Cambridge, UK: Cambridge University Press, 2003). J. Scott Turner provides a wonderful introduction to how signals are translated in the sense organs and brain into representations. His excellent book *The Tinkerer's Accomplice: How Design Emerges from Life Itself* (Cambridge, MA: Harvard University Press, 2007) offers insights into adaptive processes that few other sources can match. A brief discussion of the sensory powers of predators in comparison to animals with other diets is given in G. J. Vermeij, *Nature: An Economic History* (Princeton, NJ: Princeton University Press, 2004). The much-less-well researched topic of sensation by plants is covered in two review articles, both emphasizing the ability to sense light in contexts other than photosynthesis: H. Smith, "Phytochromes and light signal perception by plants—an emerging synthesis," *Nature* 407: 585–591 (2000); and B. L. Montgomery and J. C. Lagarias, "Phytochrome ancestry: Sensors of bilins and light," *Trends in Plant Science* 7: 357–366 (2002).

Several fine papers cover particular sensory modalities. For vision see Eric Warrant, "Vision in the dimmest habitats on earth," *Journal of Comparative Physiology* A 190: 765–789 (2004); for the sense of smell see J. L. DeBose and G. A. Nevitt, "The use of odors at different spatial scales: comparing birds and fish," *Journal of Chemical Ecology* 34: 867–881. For sound see P. Senter, "Voices of the past: a review of Paleozoic and Mesozoic animal sounds," *Historical Biology* 20: 255–287 (2008); and J. F. Gillooly and A. G. Ophir, "The energetic basis of acoustic ommunication," *Proceedings of the Royal Society* B 277: 1325–1331 (2010).

The humble origins of animal vision have recently been clarified by G. Jékely, J. Colombelli, H. Hausen, K. Guy, E. Stelzer, F. Nédérlec, and D. Arendt, "Mechanisms of phototaxis in marine zooplankton," *Nature* 456: 395–399 (2008). Sensation from the animal's point of view is richly documented with examples from the remarkable natural history of animal builders by James L. Gould and Carol G. Gould in their indispensable book, *Animal Architects: Building and the Evolution of Intelligence* (New York: Basic Books, 2007). Sensations used by snails to build shells were not discussed by the Goulds, but are well documented by A. G. Checa, A. P. Jiménez-Jiménez, and P. Rivas, "Regulation of spiral coiling in the terrestrial gastropod *Sphincterochila*: An experimental test of the road-holding model," *Journal of Morphology* 235: 249–257 (1998).

For details on experiments and insights into how sounds are transformed into a culturally evolving language, see the excellent paper by S. Kirby, H. Cornish, and K. Smith, "Cumulative cultural evolution in the laboratory: An experimental approach to the origins of structure in human languages," *Proceedings of the National Academy of Sciences of the United States of America* 105: 10681–10686 (2008). An excellent paper on the theory of how the genetic code became universal is by K. Vetsigian, C. Woese, and N. Goldenfeld, "Collective evolution of the genetic code," *Proceedings of the National Academy of Sciences of the United States of America* 103: 10696–10701 (2006).

Chapter 3: On Imperfection

The argument that antipredatory adaptation implies imperfect adaptation in both predator and prey is proposed in G. J. Vermeij, "Unsuccessful predation and evolution," *American Naturalist* 120: 701–720 (1982). The life-dinner principle was set forth by Richard Dawkins and John R. Krebs, "Arms races between and within species," *Proceedings of the Royal Society of London B* 205: 489–511 (1979). I discussed the importance of inequality in evolution in my paper "Inequality and the directionality of history," *American Naturalist* 153: 243–253 (1999); see also my book *Nature: An Economic History* (Princeton, NJ: Princeton University Press, 2004).

The inequalities in the East African acacia system are described in all their complexity by T. M. Palmer, M. L. Stanton, T. P. Young, J. R. Goheen, R. M. Pringle, and R. Karban, "Breakdown of an ant-plant mutualism follows the loss of large herbivores from an African savanna," *Science* 319: 192–195 (2008). Inequality was not the point of this paper, but as so often happens in science, data collected for unrelated reasons often turns out to be relevant in unexpected ways.

Ecological subsidies of unproductive environments by productive ones are comprehensively summarized by G. A. Polis, W. B. Anderson, and R. D. Holt, "Toward an integration of landscape and food web ecologies: The dynamics of spatially subsidized food webs," *Annual Review of Ecology and Systematics* 28: 289–316 (1997).

The natural history of animal weaponry, especially that related to mate selection, is admirably covered by Douglas J. Emlen, "The evolution of animal weapons," *Annual Review of Ecology, Evolution, and Systematics* 39: 388–413 (2008). My only substantial point of disagreement with him is his claim that weapons usually evolve to become less lethal and more slightly symbolic. In my view, animal weapons and human weapons are functional both in delivering grievous injury and in symbolically conveying power. Particularly good books on military escalation are William H. McNeill, *The Pursuit of Power: Technology, Armed Force, and Society Since A.D. 1000* (Chicago: University of Chicago Press, 1982); and M. Nincic, *The Arms Race: The Political Economy of Military Growth* (New York: Praeger, 1982).

An excellent introduction to evolutionary medicine, including strategies for fighting rapidly mutating pathogens, is R. M. Nesse and G. C. Williams, *Why We Get Sick: The New Science of Darwinian Medicine* (New York: Times Books, 1994). I have long used this book in my course on evolution as a highly accessible vehicle to connect people to the broader principles of adaptation. Strategies for slowing down arms races between pests and pesticides are outlined by R. D. Holt and M. E. Hochberg, "When is biological control evolutionarily stable (or is it)? *Ecology* 78: 1673–1683 (1997).

Chapter 4: Taming Unpredictability

The Santa Barbara workshops resulted in the publication of an excellent book with widely differing perspectives: R. D. Sagarin and

T. Taylor (eds.), *National Security: A Darwinian Approach to a Dangerous World* (Berkeley: University of California Press, 2008). The ideas on unpredictability and adaptation are developed in my chapter, "Security, unpredictability, and evolution: Policy and the history of life," pp. 25–40.

There is a very large literature on aging. I find three papers especially useful: J. R. Carey and D. S. Judge, "Lifespan in humans is self-reinforcing: A general theory of longevity," *Population and Development* 27: 411–436 (2001); H. S. Kaplan and A. J. Robson, "The emergence of humans: The coevolution of intelligence and longevity with intergenerational transfers," *Proceedings of the National Academy of Sciences of the United States of America* 99: 10221–10226 (2002); and R. Perez-Camp, M. López-Torres, S. Cadenas, C. Rojas, and G. Barja, "The rate of free radical production as a determinant of the rate of aging: Evidence from the comparative approach," *Journal of Comparative Physiology B* 168: 149–158 (1998).

The repeated tendency of human societies to ignore signs of impending doom is well documented by Jared Diamond, *Collapse: How Societies Choose to Fail or Succeed* (New York: Viking, 2005). See also John Perlin, *A Forest Journey: The Role of Wood in the Development of Civilization* (Cambridge, MA: Harvard University Press, 1989); and Daniel Hillel, *Out of the Earth: Civilization and the Life of the Soil* (Berkeley: University of California Press, 1991).

A thorough history of financial crises is given in two fine books: E. Chancellor, *Devil Take the Hindmost: A History of Financial Speculation* (New York: Plume, 1999); and N. Ferguson, *The Ascent of Money: A Financial History of the World* (New York: Penguin, 2008).

The literature on the causes of extinction is huge. See G. J. Vermeij, "Ecological avalanches and the two kinds of extinction," *Evolutionary Ecology Research* 6: 315–337 (2004); and P. D. Roopnarine, "Extinction cascades and catastrophe in ancient food webs," *Paleobiology* 32: 1–19 (2006).

Chapter 5: The Evolution of Order

Details of my work on *Littorina littorea* and *Nucella lapillus* are to be found in two papers: G. J. Vermeij, "Environmental change and the evolutionary history of the periwinkle *Littorina littorea* in North America," *Evolution* 36: 561–580 (1982); and G. J. Vermeij, "Phenotypic evolution in a poorly dispersing snail after arrival of a predator," *Nature* 299: 349–350 (1982). A great deal of experimental evidence for predator-induced changes in shell shape has accumulated. One of the earliest papers is by R. D. Appleton and A. R. Palmer, "Waterborne stimuli released by predatory crabs and damaged prey induce more predator-resistant shells in a marine gastropod," *Proceedings of the National Academy of Sciences of the United States of America* 85: 4386–4391 (1988); see also G. C. Trussell and L. D. Smith, "Induced defenses in response to an invading crab predator: An explanation for historical and geographic phenotypic change," *Proceedings of the National Academy of Sciences of the United States of America* 97: 2123–2127 (2000). Further insights into the history of the dog whelk are presented by J. A. D. Fisher, E. C. Rhile, H. Liu, and P. Petraitis, "An intertidal snail shows a dramatic size increase over the past century," *Proceedings of the National Academy of Sciences of the United States of America* 106: 5209–5212.

An excellent general review of induced form by external forces is given in D'Arcy Wentworth Thompson's classic book *On Growth and Form (Third Edition)* (Cambridge, UK: Cambridge University Press, 1942). A link among growth rate, growth direction, and outside and internal forces is explored in G. J. Vermeij, "Characters in context: Molluscan shells and the forces that mold them," *Paleobiology* 28: 41–54 (2002). Phenotypic plasticity, involving direct responses of a wide variety of organisms to the physical and chemical environment, is beautifully summarized by A. A. Agrawal, "Phenotypic plasticity in the interactions and evolution of species," *Science* 294: 321–326 (2001).

The transition from unregulated variation to strict genetic control has been widely discussed. I find the following sources particularly insightful: J. A. Doyle and L. J. Hickey, "Pollen and leaves from the mid Cretaceous Potomac Group and their bearing on early angiosperm evolution," pp. 139–206 in C. B. Beck (ed.), *Origin and Early Evolution of Angiosperms* (New York: Columbia University Press, 1976); C. H. Waddington, *New Patterns in Genetics and Development* (New York: Columbia University Press, 1962); M. Pigliucci, "Characters and environments," pp. 363–388 in G. P. Wagner (ed.), *The Character Concept in Evolutionary Biology* (San Diego: Academic Press, 2001); A. R. Palmer, "Symmetry breaking and the evolution of development," *Science* 306: 828–833 (2004); M. Webster, "A Cambrian peak in morphological variation within trilobite species," *Science* 317: 499–502 (2007). The effect of learning on the acquisition of genetic traits is explored by E. Danchin, L.-A. Giraldeau, T. J. Valone, and R. H. Wagner, "Public information: From noisy neighbors to cultural evolution," *Science* 305: 487–491 (2004).

The case of a weedy, unregulated body form at the base of major evolutionary branches is most forcefully made by Stephen Jay Gould in his *Ontogeny and Phylogeny* (Cambridge, MA: Belknap Press of Harvard University, 1977). A case for sloppier shell construction at higher latitudes has been made by A. H. Clarke, Jr., "Polymorphisms in marine mollusks and biome development," *Smithsonian Contributions to Zoology* 274: 1–14 (1978).

The themes of flexibility and innovation are persuasively developed by J. L. Gould and C. G. Gould, *Animal Architects: Building and the Evolution of Intelligence* (New York: Basic Books, 2007).

Chapter 6: The Complexity of Life and the Origin of Meaning

Stuart A. Kauffman's *Reinventing the Sacred: A New View of Science, Reason, and Religion* (New York: Basic Books, 2008) is an extended treatise on emergence. Besides offering a fine discussion of the origins of ethics, the book presents the most coherent and credible hypothesis for life's origin that I have read. See also his earlier *Investigations* (Oxford, UK: Oxford University Press, 2000), in which networks and phase transitions are particularly well described. Also important is the paper by M. A. Nowak and H. Ohtsuki, "Prevolutionary dynamics and the origin of evolution," *Proceedings of the National Academy of Sciences of the United States of America* 105: 14924–14927 (2008), in which the authors argue—correctly, I believe—that selection preceded the evolution of replication.

My perspective on the evolution of the genetic code follows that of Carl Woese, "Interpreting the universal phylogenetic tree," *Proceedings of the National Academy of Sciences of the United States of America* 97: 8392–8396 (2000). See also the suggested readings for chapter 2. The evolution of sex and recombination has been the subject of prolific speculation. One of the most illuminating papers in this area is Wolfgang Sterrer's "On the origin of sex as vaccination," *Journal of Theoretical Biology* 216: 387–396 (2002). Another very useful paper on the related topic of the origin of multicellularity is R. K. Grosberg and R. R. Strathmann, "The evolution of multicellularity: A minor major transition," *Annual Reviews of Ecology, Evolution, and Systematics* 38: 621–654 (2007). Hybridization as a source of innovation is discussed by J. Mallet, "Hybrid speciation," *Nature* 446: 279–283 (2007).

The many advantages of group life and the evolutionary pathways toward the establishment of colonial and social animals have been the subject of numerous important works. For the advantages of group hunting see C. Packer and L. Ruttan, "The evolution of cooperative hunting," *American Naturalist* 132: 159–198 (1988); and E. C. Yip, S. Powers, and L. Avilés, "Cooperative capture of large prey solves scaling challenge faced by spider societies," *Proceedings of the National Academy of Sciences of the United States of America* 105:

11818–11822 (2008). Alternative viewpoints on the evolution of caste societies in insects can be found in E. O. Wilson and B. Hölldobler, "Eusociality: Origin and consequences," *Proceedings of the National Academy of Sciences of the United States of America* 102: 13367–13371 (2005); and H. K. Reeve and B. Hölldobler, "The emergence of a superorganism through intergroup competition," *Proceedings of the National Academy of Sciences of the United States of America* 104: 6736–6740 (2007). Wilson and Hölldobler hold that close kinship is the consequence of the evolution of insect sociality, a view I prefer over that of Reeve and Hölldobler, who support the older view championed by W. D. Hamilton that close kinship is responsible for the peculiar social organization of ants, bees, and other Hymenoptera.

The most incisive work on colonial marine animals has been done by Jeremy B. C. Jackson and his colleagues. See, for example, J. B. C. Jackson, "Biological determinants of present and past sessile animal distributions," pp. 39–120 in M. J. S. Tevesz and P. L. McCall (eds.), *Biotic Interactions in Recent and Fossil Benthic Communities* (New York: Plenum, 1983); J. B. C. Jackson and A. G. Coates, "Life cycles and evolution of clonal (modular) animals," *Philosophical Transactions of the Royal Society of London B* 313: 7–22 (1986); and A. G. Coates and J. B. C. Jackson, "Morphological themes in the evolution of clonal and aclonal invertebrates," pp. 67–106 in J. B. C. Jackson, L. W. Buss, and R. E. Cook (eds.), *Population Biology and Evolution of Clonal Organisms* (New Haven, CT: Yale University Press, 1985).

Useful sources for the evolution of human altruism, cooperation, and sociality include L. Dugatkin, *Cheating Monkeys and Citizen Bees: The Nature of Cooperation in Animals and Humans* (New York: Free Press, 1999); P. Seabright, *The Company of Strangers: A Natural History of Economic Life* (Princeton, NJ: Princeton University Press, 2004); M. D. Hauser, *Moral Minds: How Nature Designed Our Universal Sense of Right and Wrong* (New York: HarperCollins, 2006);

S. Bowles, "Conflict, altruism's midwife," *Nature* 456: 326–327 (2008); and S. Bowles, "Policies designed for self-interested citizens may undermine 'the moral sentiments': Evidence from economic experiments," *Science* 320: 1605–1609 (2008). The opinions I express in chapter 8 take these and many other sources into account but do not precisely reflect any of them.

Also very enlightening is the paper by J. Henrich, J. Ensminger, R. McElreath, A. Barr, C. Barrett, A. Bolyanatz, J. Camilo Cardenas, M. Gurven, E. Gwako, N. Henrich, C. Lesorogol, F. Marlowe, D. Tracer, and J. Riker, "Markets, religion, community size, and the evolution of fairness and punishment," *Science* 327: 1480–1484 (2010).

Parallels between the human and nonhuman economies of life are set out in G. J. Vermeij, "Comparative economics, evolution and the modern economy," *Journal of Bioeconomics* 11: 105–134 (2009). In this paper I make the case that humans are more potent than, but not fundamentally different from, other organisms in their economic relationships. See also G. J. Vermeij, *Nature: An Economic History* (Princeton, NJ: Princeton University Press, 2004).

Although I have not delved deeply into the philosophical literature on intentionality, Kauffman's *Reinventing the Sacred* (see above) and David L. Hull's *Science and Selection: Essays on Biological Evolution and the Philosophy of Science* (Cambridge, UK: Cambridge University Press, 2001) have helped shape my thoughts on this subject. See also Daniel C. Dennett's *Darwin's Dangerous Idea: Evolution and the Meaning of Life* (New York: Simon and Schuster, 1995).

For goal-directed behavior and the evolution of intentionality in animals see Frans B. M. de Waal's *Good Natured: The Origins of Right and Wrong in Humans and Other Animals* (Cambridge, MA: Harvard University Press, 1996); N. J. Emery and N. S. Clayton, "The mentality of crows: Convergent evolution of intelligence in corvids and apes," *Science* 306: 1903–1907 (2004); A. B. Butler and R. M.

J. Cotterill, "Mammalian and avian neuroanatomy and the question of consciousness in birds," *Biological Bulletin* 211: 106–127 (2006); C. R. Raby, D. M. Alexis, A. Dickinson, and N. S. Clayton, "Planning for the future by Western scrub-jays," *Nature* 445: 919–921 (2007); J. L. Gould and C. G. Gould, *Animal Architects: Building and the Evolution of Intelligence* (New York: Basic Books, 2007); J. A. Mather, "Cephalopod consciousness: Behavioural evidence," *Consciousness and Cognition* 17: 37–48 (2008); and especially J. Scott Turner, *The Tinkerer's Accomplice: How Design Emerges from Life Itself* (Cambridge, MA: Harvard University Press, 2007).

Excellent summaries of the cultural evolution of tool use and language, from simple to compound structures, are found in S. H. Ambrose, "Paleolithic technology and human evolution," *Science* 291: 1748–1753 (2001); and S. Mithen, *The Singing Neanderthals: The Origins of Music, Language, Mind and Body* (Cambridge, MA: Harvard University Press, 2006). For the times of behavioral transitions in human evolution see Z. Jacobs, R. G. Roberts, R. F. Galbraith, H. J. Deacon, R. Grün, A. Mackay, P. Mitchell, R. Vogelsang, and L. Wadley, "Ages for the Middle Stone Age of southern Africa: Implications for human behavior and dispersal," *Science* 322: 733–735 (2008).

Countless tracts spew misleading and incendiary drivel about the meaninglessness of life without a god. An especially popular representative example, from which I quote, is Rick Warren's *The Purpose Driven Life: What on Earth Am I Here For?* (Grand Rapids, MI: Zondervan, 2002). Stuart A. Kauffman's *Reinventing the Sacred* (see above) presents a welcome antidote.

Chapter 7: The Secrets of Grass: Interdependence and Its Discontents

The most comprehensive account of the evolution of grazing mammals is Christine M. Janis's "An evolutionary history of browsing and grazing ungulates," pp. 21–45 in J. J. Gordon and R. H. T. Prins (eds.), *The Ecology of Browsing and Grazing* (Berlin: Springer, 2008). The characteristics of grasses that make these plants highly adapted to intense grazing are discussed by D. F. Owen, "How plants may benefit from the animals that eat them," *Oikos* 35: 230–235 (1980); and C. M. Herrera, "Grasses, grazers, mutualism, and coevolution: A comment," *Oikos* 38: 254–258 (1981). Important papers on the evolution of grasses include K. Bremer, "Gondwanan evolution of the grass alliance of families (Poales)," *Evolution* 56: 1374–1387 (2002); C. A. E. Strömberg, "Decoupled taxonomic radiation and ecological expansion of open habitat grasses in the Cenozoic of North America," *Proceedings of the National Academy of Sciences of the United States of America* 102: 11980–11984 (2005); and A. Vicentini, C. Barber, S. S. Aliscioni, L. M. Giussani, and E. A. Kellogg, "The age of the grasses and clusters of origins of C4 photosynthesis," *Global Change Biology* 14: 2963–2977 (2008).

Important studies of plant-eating dinosaurs' diet and physiology include J. M. Lehman and H. N. Woodward, "Modeling growth rates for sauropod dinosaurs," *Paleobiology* 34: 264–281 (2008); and J. Hummel, C. T. Gee, K.-H. Südekum, P. M. Sander, G. Nogge, and M. Clauss, "*In vitro* digestibility of fern and gymnosperm foliage: Implications for sauropod feeding ecology and diet selection," *Proceedings of the Royal Society B* 275: 1015–1021 (2008). The most succinct argument that dinosaurs did not normally feed high in the trees is in a comment by R. S. Seymour, "Sauropods kept their heads down," *Science* 323: 1671 (2009). For the use of the neck in sexual selection in the giraffe, see R. E. Simmons and L. Scheepers, "Winning by a neck: Sexual selection in the evolution of giraffe," *American Naturalist* 148: 771–786 (1996).

The distribution of grasses on continents but not islands is documented by E. G. Leigh, Jr., A. Hladik, C. L. Hladik, and A. Jolly,

"The biogeography of large islands, or how does the size of the ecological theater affect the evolutionary play?", Revue d'Ecologie (La Terre et La Vie) 62: 105–168 (2007). This paper also strongly emphasizes the importance of interdependencies in ecosystems.

The history of herbivory in general is given in chapter 10 of G. J. Vermeij, *Nature: An Economic History* (Princeton, NJ: Princeton University Press, 2004). The suggestion there that network venation provides redundancy against herbivores is supported by L. Sack, E. M. Dietrich, C. M. Streeter, D. Sánchez-Gómez, and N. M. Holbrook, "Leaf palmate venation and vascular redundancy confer tolerance of hydraulic disruption," *Proceedings of the National Academy of Sciences of the United States of America* 105: 1567–1572 (2008). Herbivory through time from the perspective of insects is summarized by C. C. Labandeira, "The four phases of plant-arthropod associations in deep time," *Geologica Acta* 4: 409–438 (2006).

For the marine record of plants and herbivory see G. J. Vermeij and D. R. Lindberg, "Delayed herbivory and the assembly of marine benthic ecosystems," *Paleobiology* 26: 419–430 (2000).

The history of plant productivity on land is receiving increased attention. The most important paper to date is by C. K. Boyce, T. J. Brodribb, T. S. Feild, and M. A. Zwieniecki, "Angiosperm leaf vein evolution was physiologically and environmentally transformative," *Proceedings of the Royal Society B* 276: 1771–1776 (2009). For the history and effects of agriculture by nonhumans see H. Hata and M. Kato, "A novel obligate cultivation mutualism between damselfish and *Polysiphonia* algae," *Biology Letters* 2: 593–596 (2006); T. R. Schultz and S. G. Brady, "Major evolutionary transitions in ant agriculture," *Proceedings of the National Academy of Sciences of the United States of America* 105: 5435–5440 (2008); and U. G. Mueller and N. Gerardo, "Fungus-farming insects: Multiple origins and diverse evolutionary histories," *Proceedings of the National Academy of*

Sciences of the United States of America 99: 15247–15249 (2002). Good summaries of the development of human agriculture are in D. R. Harris, "Climatic change and the beginnings of agriculture, the case of the Younger Dryas," pp. 379–394 in L. J. Rothschild and A. Lister (eds.), *Evolution on Planet Earth: The Impact of the Physical Environment* (Amsterdam: Academic Press, 2003); D. R. Piperno, "The origins of plant cultivation and domestication in the Neotropics: A behavioral ecological perspective," pp. 137–166 in D. Kennett and B. Winterhalder (eds.), *Behavioral Ecology and the Transition to Agriculture* (Berkeley: University of California Press, 2006); and M. D. Purugganan and D. Q. Fuller, "The nature of selection during plant domestication," *Nature* 457: 843–848 (2009). I disagree with Harris and Piperno about the climatic trigger for human agriculture, but these authors summarize the best available chronology of events, supplemented with new data and evolutionary insights by Purugganan and Fuller.

Disruptions of highly interdependent ecosystems have been documented in many papers, including J. B. C. Jackson, M. X. Kirby, W. H. Berger, K. A. Bjorndal, L. W. Botsford, B. J. Bourque, R. H. Bradbury, R. Cooke, J. Erlandson, J. A. Estes, T. P. Hughes, S. Kidwell, C. B. Lang, H. S. Lenihan, J. M. Pandolfi, C. H. Peterson, R. S. Steneck, M. J. Tegner, and R. R. Warner, "Historical overfishing and the recent collapse of coastal ecosystems," *Science* 293: 629–638 (2001); W. J. Ripple, E. J. Larsen, R. A. Renkin, and D. W. Smith, "Trophic cascades among wolves, elk and aspen on Yellowstone National Park's northern range," *Biological Conservation* 102: 227–234 (2001); B. Worm, E. B. Barbier, N. Beaumont, J. E. Duffy, C. Folke, B. S. Halpern, J. B. C. Jackson, H. K. Lotze, F. Micheli, S. R. Palumbi, E. Sala, K. A. Selkoe, J. J. Stachowicz, and R. Watson, "Impact of biodiversity loss on ocean ecosystem services," *Science* 314: 787–790 (2007); G. J. Vermeij, "Ecological avalanches and the two kinds of extinction," *Evolutionary Ecology Research* 6: 315–337 (2004). Ecological dead zones and the nutrient enrichment that causes them are reviewed in R. J. Diaz and R. Rosenberg, "Spreading dead zones and consequences for marine ecosystems," *Science* 321: 949–952 (2008).

Chapter 8: Nature's Housing Market, or Why Nothing Happens in Isolation

A review of many animal groups that occupy discarded molluscan shells is given in chapter 8 of G. J. Vermeij, *Evolution and Escalation: An Ecological History of Life* (Princeton, NJ: Princeton University Press, 1987). Hermit-crab shells as microcosms are thoroughly summarized by J. D. Williams and J. D. McDermott, "Hermit crab biocoenoses: A worldwide review of the diversity and natural history of hermit crab associations," *Journal of Experimental Marine Biology and Ecology* 305: 1–128 (2004). The relationships among sipunculans, shells, and corals, including their fossil record, are given in G. A. Gill and A. G. Coates, "Mobility, growth patterns and substrate in some fossil and Recent corals," *Lethaia* 10: 119–134 (1917); and J. Stolarski, H. Zibrowius, and H. Loser, "Antiquity of the scleractinian-sipunculan symbiosis," *Acta Palaeontologica Polonica* 46: 309–330 (2001). Reports of Ordovician-age trilobites in the shells of cephalopod molluscs appear in R. A. Davis, R. H. B. Fraaye, and C. H. Holland, "Trilobites within nautiloid cephalopods," *Lethaia* 34: 37–45 (2001).

Many studies of competition for, and patterns of occupation of, shells by hermit crabs have been done. Two of the best are by my former student Mark D. Bertness, "Patterns and plasticity in tropical hermit crab growth and reproduction," *American Naturalist* 117: 754–763 (1981); and "Conflicting advantages in resource utilization: The hermit crab housing dilemma," *American Naturalist* 118: 432–437 (1981). The *Stylobates* anomaly was clarified in an excellent paper by D. F. Dunn, D. M. Devaney, and B. Roth, "*Stylobates*: A shell-forming sea anemone (Coelenterata, Anthozoa, Actiniidae)," *Pacific Science* 34: 379–388 (1980). Experiments on how the hermit-crab body form is modified according to the shape of the shell cavity were conducted by N. W. Blackstone, "The effects of shell size and shape on growth and form in the hermit crab *Pagurus longicarpus*," *Biological Bulletin* 168: 75–90 (1985).

R. B. McLean's account of hermit crabs waiting at predation sites is the earliest of several known examples of this phenomenon; see his "Direct shell acquisition by hermit crabs from gastropods," *Experientia* 30: 206–208 (1974). See also the excellent paper by T. P. Wilber and W. F. Herrnkind, "Predaceous gastropods regulate new-shell supply to salt marsh hermit crabs," *Marine Biology* 79: 145–150 (1984).

The central role of synergy in complexity is fully set out in Peter Corning's *Holistic Darwinism: Synergy, Cybernetics and the Bioeconomics of Evolution* (Chicago: University of Chicago Press, 2005). See also my book *Nature: An Economic History* (Princeton, NJ: Princeton University Press, 2004).

Chapter 9: Dispatches from a Warmer World

An outstanding summary of the effects of temperature on the physical properties of materials and on biological phenomena is Mark W. Denny's *Air and Water: The Biology and Physics of Life's Media* (Princeton, NJ: Princeton University Press, 1993). A good summary of life in the cold is given by Andrew Clarke in "Life in cold water: The physiological ecology of polar marine ectotherms," *Oceanography and Marine Biology Annual Reviews* 21: 341–453 (1983).

The best account of tropical forests I have read is by my friend Egbert G. Leigh, Jr., *Tropical Forest Ecology: A View from Barro Colorado Island* (New York: Oxford University Press, 1999). An excellent comparative account of the biology of lianas is given by S. A. Schnitzer,

"A mechanistic explanation for global patterns of liana abundance and distribution," *American Naturalist* 166: 262–276 (2005).

Detailed comparisons of the tropics and colder zones with respect to adaptive potential appear in G. J. Vermeij, *Biogeography and Adaptation: Patterns of Marine Life* (Cambridge, MA: Harvard University Press, 1978); G. J. Vermeij, *A Natural History of Shells* (Princeton, NJ: Princeton University Press, 1993); P. D. Coley and T. M. Aide, "Comparison of herbivory and plant defenses in temperate and tropical broad-leaved forests," pp. 25–49 in P. W. Price, T. M. Lewinsohn, G. W. Fernandes, and W. B. Benson (eds.), *Plant-Animal Interactions: Evolutionary Ecology in Tropical and Temperate Regions* (New York: John Wiley and Sons, 1991); and G. J. Vermeij, "Temperature, tectonics, and evolution," pp. 209–232 in L. J. Rothschild and A. Lister (eds.), *Evolution on Planet Earth: The Impact of the Physical Environment* (San Diego: Academic Press, 2003).

Evidence for a link between warming and diversification is given by J. L. Cornette, B. S. Lieberman, and R. H. Goldstein, "Documenting a significant relationship between macroevolutionary origination and Phanerozoic pCO_2 levels," *Proceedings of the National Academy of Sciences of the United States of America* 99: 7832–7835 (2002).

The physiology of warm-blooded animals has been thoroughly reviewed by John A. Ruben, "The evolution of endothermy in mammals and birds: From physiology to fossils," *Reviews of Physiology* 57: 69–95 (1995); Bernd Heinrich, *The Hot-Blooded Insects: Strategies and Mechanisms of Thermoregulation* (Cambridge, MA: Harvard University Press, 1993); K. A. Dickson and J. B. Graham, "Evolution and consequences of endothermy in fishes," *Physiological and Biochemical Zoology* 77: 998–1018 (2004). For plants see R. S. Seymour, C. R. White, and M. Gibernau, "Heat reward for insect pollinators," *Nature* 426: 243–244 (2003).

A full account of the labral tooth and its evolution in predatory snails is given in G. J. Vermeij, "Innovation and evolution at the edge: Origins and fates of gastropods with a labral tooth," *Biological Journal of the Linnean Society* 72: 461–508 (2001).

A good summary of the latitudinal gradient in diversity is given in M. L. Rosenzweig, *Species Diversity in Space and Time* (Cambridge, UK: Cambridge University Press, 1995); and H. Hillebrand, "On the generality of the latitudinal diversity gradient," *American Naturalist* 163: 192–211 (2004). The role of selection in creating isolation among populations is discussed by M. J. West-Eberhard, "Sexual selection, social competition, and speciation," *Quarterly Review of Biology* 58: 155–183 (1983). A microbial perspective on this issue is given in several fine papers, including M. Llewellyn and L. L. Rainey, "Ecological constraints on diversification in a model adaptive radiation," *Nature* 431: 984–988 (2004); and J. R. Meyer and R. Kassen, "The effects of competition and predation on diversification in a model adaptive radiation," *Nature* 446: 432–435 (2007). The high diversity of tropical trees, with emphasis on the role of enemies, is the subject of the excellent paper by E. G. Leigh, Jr., P. Davidar, C. W. Dick, J.-P. Puyravaud, J. Terborgh, H. ter Steege, and S. J. Wright, "Why do some tropical forests have so many species of trees?" *Biotropica* 36: 447–473 (2004).

Linguistic diversity and isolation are discussed by M. Pagel and R. Mace, "The cultural wealth of nations," *Nature* 428: 275–278 (2004). A general consideration of diversity and its explanation appears in my paper "From phenomenology to first principles: Toward a theory of diversity," *Proceedings of the California Academy of Sciences* 56, Supplement I (2): 12–23 (2005).

Chapter 10: The Search for Sources and Sinks

African origins for hominids and of the modern human species are generally accepted; see, for example, A. R. Templeton, "Out of Africa again and again," *Nature* 416: 45–51 (2002). For a contrarian view, see R. Dennell and W. Roebroeks, "An Asian perspective on early human dispersal," *Nature* 438: 1099–1104 (2005).

Potential effects of Pacific predators in the Caribbean have been discussed by P. W. Glynn, "Coral communities and their modifications related to past and prospective Central American seaways," *Advances in Marine Biology* 19: 91–132 (1982); I. Rubinoff, "Central American sea-level canal: Possible biological effects," *Science* 161: 857–861 (1968); and I. Rubinoff and C. Kropach, "Differential reactions of Atlantic and Pacific predators to sea snakes," *Nature* 228: 1288–1290 (1970).

The link between competitiveness and geographic location and size of ecosystems was already understood by Charles Darwin in his justifiably famous book *The Origin of Species by Natural Selection or The Preservation of Favoured Races in the Struggle for Life* (London: John Murray, 1859). See also J. C. Briggs, "Dispersal of tropical marine shore animals: Coriolis parameters or competition?" *Nature* 216: 350 (1966); G. J. Vermeij, "When biotas meet: Understanding biotic interchange," *Science* 253: 1099–1104 (1991); and G. J. Vermeij, "Invasion as expectation: A historical fact of life," pp. 315–339 in D. F. Sax, J. J. Stachowicz, and S. D. Gaines (eds.), *Species Invasions: Insights into Ecology, Evolution, and Biogeography* (Sunderland, MA: Sinauer Associates, 2005). Though now somewhat outdated and not particularly well written, the best single source for the trans-Suez invasions is D. F. Por, *The Legacy of Tethys: An Aquatic Biogeography of the Levant* (Dordrecht: Kluwer, 1989). For the trans-Arctic interchange see G. J. Vermeij, "Anatomy of an invasion: The trans-Arctic interchange," *Paleobiology* 17: 281–307 (1991); and G. J. Vermeij and P. D. Roopnarine, "The coming Arctic invasion," *Science* 321: 780–781 (2008). The best summary of the Great American interchange is S. D. Webb and A. Rancy, "Late Cenozoic evolution of the neotropical mammal fauna," pp. 335–358 in J. B. C. Jackson, A. F. Budd, and A. G. Coates (Eds.), *Evolution and Environment in Tropical*

America (Chicago: University of Chicago Press, 1996); see also papers in the excellent book edited by F. G. Stehli and S. D. Webb, *The Great American Biotic Interchange* (New York: Plenum, 1985).

Among the important papers on the marine-biological history of tropical America, including patterns of extinction, are the following: W. P. Woodring, "The Panama land bridge as a sea barrier," *American Philosophical Society Proceedings* 110: 425–433 (1966); A. O'Dea, J. B. C. Jackson, H. Fortunato, J. T. Smith, L. D'Cruz, K. G. Johnson, and J. A. Todd, "Environmental change preceded Caribbean extinction by 2 million years," *Proceedings of the National Academy of Science of the United States of America* 104: 5501–5506 (2007); and P. Molnar, "Closing of the Central American Seaway and the ice age: A critical review," *Paleoceanography* 23, Article PA2201, DOI: 10.1029/2007PA2001574 (2008). Although O'Dea and colleagues give the most detailed environmental history of the Panama seaway and its vicinity published to date, I disagree with their contention that extinction followed long after the major productivity decrease in the Caribbean; see B. Landau, C. M. da Silva, and G. J. Vermeij, "Pacific elements in the Caribbean Neogene gastropod fauna: The source-sink model, larval development, disappearance, and faunal units," *Bulletin de la Société Géologique de France* 180: 249–258 (2009).

The evolutionary significance of source regions and populations has been examined by H. R. Pulliam, "Sources, sinks, and population regulation," *American Naturalist* 132: 652–661 (1988); R. D. Holt, "Demographic constraints in evolution: Towards unifying the evolutionary theories of senescence and niche conservatism," *Evolutionary Ecology* 10: 1–11 (1996); and G. J. Vermeij and G. P. Dietl, "Majority rule: Adaptation and the long-term dynamics of species," *Paleobiology* 32: 273–278 (2006).

Classic papers on punctuated equilibria include those of N. Eldredge, "The allopatric model and phylogeny in Paleozoic invertebrates," *Evolution* 25: 156–166 (1971); and N. Eldredge and S. J. Gould, "Punctuated equilibria: An alternative to phyletic gradualism," pp. 82–115 in T. J. M. Schopf (ed.), *Models in Paleobiology* (San Francisco: Freeman, Cooper, and Co., 1972). These papers were preceded by H. J. MacGillavry, "Modes of evolution mainly among marine invertebrates: An observational approach," *Bijdragen tot de Dierkunde* 38: 69–74 (1968). For Gould's last revisions of ideas on punctuated equilibria see S. J. Gould, *The Structure of Evolutionary Theory* (Cambridge, MA: Belknap Press of Harvard University, 2002).

Differences between island and continental ecosystems and their implications for human-caused habitat fragmentation are discussed by E. G. Leigh, Jr., G. J. Vermeij, and M. Wikelski, "What do human economies, large islands and forest fragments tell us about the factors limiting ecosystem evolution?", *Journal of Evolutionary Biology* 22: 1–12 (2009). For a marine evolutionary perspective on islands see G. J. Vermeij, "Island life: A view from the sea," pp. 239–254 in M. V. Lomolino and L. R. Heaney (eds.), *Frontiers of Biogeography: New Directions in the Geography of Nature* (Sunderland, MA: Sinauer Associates, 2004).

A comprehensive paper on extinct mammals on Pleistocene Mediterranean islands is by P. Raia and S. Meiri, "The island rule in large mammals: Paleontology meets ecology," *Evolution* 60: 1731–1742 (2006). Two good sources (among many) on *Homo floresiensis* are M. J. Morwood, R. P. Soedjono, R. G. Roberts, T. Sutikna, C. S. M. Turney, K. E. Westaway, W. J. Rink, J.-X. Zhao, J. D. van den Bergh, R. Awe Due, D. R. Hobbs, M. W. Moore, M. I. Bird, and L. K. Fifield, "Archaeology and age of a new hominin from Flores, eastern Indonesia," *Nature* 431: 1087–1091 (2004); and W. L. Jungers, W. E. H. Harcourt-Smith, R. E. Wunderlich, M. W. Tocheri, S. G. Larson, T. Sutikna, R. Awe Due, and M. J. Morwood, "The foot of *Homo floresiensis*," *Nature* 459: 81–84 (2009).

Chapter 11: Invaders, Incumbents, and a Changing of the Guard

The rarity of insects in the sea and the general role of incumbency in evolution are treated in G. J. Vermeij and R. Dudley, "Why are there so few evolutionary transitions between aquatic and terrestrial ecosystems?", *Biological Journal of the Linnean Society* 70: 541–554 (2000); see also M. L. Rosenzweig and R. D. McCord, "Incumbent replacement: Evidence of long-term evolutionary progress," *Paleobiology* 17: 202–213 (1991). For a provocative account of the colonization of the land by marine and freshwater organisms in the distant past see G. J. Retallack, "Ordovician life on land and Early Paleozoic global change," pp. 21–45 in R. A. Gastaldo and W. A. DiMichele (eds.), *Phanerozoic Terrestrial Ecosystems* (Paleontological Society Papers 6, 2000). Valuable physiological perspectives on this transition are offered by J. B. Graham and H. J. Lee, "Breathing air in air: In what ways might extant amphibious fish physiology relate to prevailing concepts about early tetrapods, the evolution of vertebrate air breathing, and the vertebrate land transition?" *Physiological and Biochemical Zoology* 77: 720–731 (2004). Transitions from sea or land to freshwater are dealt with in important papers by C. D. K. Cook, "The number and kinds of embryo-bearing plants which have become aquatic: A survey," *Perspectives in Plant Ecology, Evolution and Systematics* 2: 79–102 (1999); F. P. Wesselingh, "Long-lived lake molluscs as island faunas: A bivalve perspective," pp. 275–314 in W. Renema (ed.), *Biogeography, Time, and Place: Distributions, Barriers, and Islands* (Berlin: Springer, 2007); and E. E. Strong, O. Gargominy, W. F. Ponder, and P. Bouchet, "Global diversity of gastropods (Gastropoda; Mollusca) in freshwater," *Hydrobiologia* 595: 149–166 (2008). The early history of animal communities on land is expertly summarized by W. A. Shear and P. A. Selden, "Rustling in the undergrowth: Animals in early terrestrial ecosystems," pp. 29–51 in P. G. Gensel and D. Edwards (eds.), *Plants Invade the Land:*

Evolutionary and Environmental Perspectives (New York: Columbia University Press, 2001).

Well-documented land-to-sea transitions in vertebrates are described for whales by P. D. Gingerich, N. A. Wells, D. H. Russell, and S. M. I. Shah, "Origin of whales in epicontinental remnant seas: New evidence from the Early Eocene of Pakistan," *Science* 220: 403–406 (1983); and J. G. M. Thewissen, L. N. Cooper, M. T. Clementz, S. Bajpai, and B. N. Tiwari, "Whales originated from aquatic artiodactyls in the Eocene epoch of India," *Nature* 450: 1190–1194 (2007); and for mosasaurs by M. W. Caldwell and A. Palci, "A new basal mosasauroid from the Cenomanian (U. Cretaceous) of Slovenia with a review of mosasauroid phylogeny and evolution," *Journal of Vertebrate Paleontology* 27: 863–880 (2007). The exceptional vulnerability of large predators to extinction during times of crisis has been widely documented. See B. Van Valkenburgh, X. Wang, and J. Damuth, "Cope's Rule, hypercarnivory, and extinction in North American canids," *Science* 306: 101–104 (2004); and L. Van Valen, "Group selection, sex, and fossils," *Evolution* 29: 87–93 (1975).

Chapter 12: The Arrow of Time and the Struggle for Life

The case for historical contingency has been strongly put by S. J. Gould, *The Structure of Evolutionary Theory* (Cambridge, MA: Belknap Press of Harvard University, 2002); E. J. Chaisson, *Cosmic Evolution: The Rise of Complexity in Nature* (Cambridge, MA: Harvard University Press, 2001); and S. A. Kauffman, *Investigations* (Oxford, UK: Oxford University Press, 2000). A more deterministic view is taken by Simon Conway Morris, *Life's Solution: Inevitable Humans in a Lonely Universe* (Cambridge, UK: Cambridge University Press, 2003). He particularly emphasizes convergence. For other discussions of contingency see C. de Duve, *Singularities: Landmarks on the Pathways of Life* (Cambridge, UK: Cambridge University Press, 2005); and G. J. Vermeij, "Historical contingency and the purported uniqueness of evolutionary innovations," *Proceedings of the National Academy of Sciences of the United States of America* 103: 1804–1809 (2006).

Viewpoints about whether complex features like eyes, limbs, and mineralized skeletons arose only once with the evolution of essential genes or much more often when the potential of these genes is realized are given by D. K. Jacobs, C. G. Wray, C. J. Wedeen, R. Kostriken, R. DeSalle, J. L. Staton, R. D. Gates, and D. R. Lindberg, "Molluscan *engrailed* expression, serial organization, and shell evolution," *Evolution & Development* 2: 340–347 (2000); and L. von Salvini-Plawen, "Photoreception and the phylogenetic evolution of photoreceptors (with special reference to Mollusca)," *American Malacological Bulletin* 26: 83–100 (2008).

The best documented example of parallel evolution I have come across is the case of the West Indian *Anolis* lizards: J. B. Losos, T. R. Jackman, A. Larson, K. de Queiroz, and L. Rodríguez-Schettino, "Contingency and determinism in replicated adaptive radiations of island lizards," *Science* 279: 2115–2118 (1998). For the case of the saber-tooth cats and other saber-tooth vertebrates see G. J. Slater and B. Van Valkenburgh, "Long in the tooth: Evolution of saber-tooth cat cranial shape," *Paleobiology* 34: 403–419 (2008).

The burgeoning study of development from an evolutionary perspective ("evo-devo") has already yielded some splendid publications. I strongly recommend R. A. Raff, *The Shape of Life: Genes, Development, and the Evolution of Animal Form* (Chicago: University of Chicago Press, 1996); S. B. Carroll, *Endless Forms Most Beautiful: The New Science of Evo Devo and the Making of the Animal Kingdom* (New York: W. W. Norton, 2005); and M. W. Kirschner and J. C. Gerhart, *The Plausibility of Life: Resolving Darwin's Dilemma* (New Haven, CT: Yale University Press, 2005). For discussions of arthropod appendages, including their genetic basis and history, see N. Shubin, A. Tabin, and S. Carroll, "Deep homology and the origins of evolutionary novelty," *Nature* 447: 818–823 (2008); and S. J. Adamowicz, A. Purvis, and M. A. Wills, "Increasing morphological complexity in multiple parallel lineages of the Crustacea," *Proceedings of the National Academy of Sciences of the United States of America* 105: 4786–4791 (2008). I explored versatility in the article "Adaptation, versatility, and evolution," *Systematic Zoology* 20: 466–477 (1973). For versatility in branching and leaf venation in plants, see the excellent paper by C. K. Boyce and A. H. Knoll, "Evolution of developmental potential and the multiple independent origins of leaves in Paleozoic vascular plants," *Paleobiology* 28: 70–100 (2002).

For gene duplication and divergence see two papers by M. Lynch and J. S. Conery: "The evolutionary fate and consequences of duplicate genes," *Science* 290: 1151–1155 (2000); and "The origins of genome complexity," *Science* 302: 1402–1404 (2003).

Work on neck vertebrae is beautifully described by F. Galis, "Why do almost all mammals have seven cervical vertebrae?" *Journal of Experimental Zoology* 285: 19–26 (1999); and F. Galis and J. A. J. Metz, "Evolutionary novelties: The making and breaking of pleiotropic constraints," *Integrative and Comparative Biology* 47: 409–419 (2007).

Documentation of resource depletion in forests is provided by D. A. Wardle, L. R. Walker, and R. D. Bardgett, "Ecosystem properties and forest decline in contrasting long-term chronosequences," *Science* 305: 509–513 (2004).

Increased control of resources by life itself over time is the subject of many important papers, among them A. G. Fischer, "Biological innovations and the sedimentary record," pp. 145–157 in H. D. Holland and A. F. Trendall (eds.), *Patterns of Change in Earth Evolution* (Berlin: Springer, 1984); T. M. Lenton, "The role of land plants, phosphorus weathering and fire in the rise and regulation of atmospheric oxygen," *Global Change Biology* 7: 613–629 (2001); A. H.

Knoll, I. J. Fairchild, and K. Swett, "Calcified microbes in Neoproterozoic carbonates: Implications for our understanding of the Proterozoic/Cambrian transition," *Palaios* 8: 512–525 (1993); and A. Ridgwell and R. E. Zeebe, "The role of the global carbonate cycle in the regulation and evolution of the Earth system," *Earth and Planetary Science Letters* 234: 299–315 (2005).

Stepwise increases in oxygen and productivity are further documented by F. Masuda and Y. Ezaki, "A great revolution of the Earth-surface environment: Linking the bio-invasion onto the land with the Ordovician radiation of marine organisms," *Paleontological Research* 13: 3–8 (2009); and L. P. Knauth and M. J. Kennedy, "The Late Precambrian greening of the Earth," *Nature* 460: 728–732 (2009).

A general history of predation, together with arguments for directionality in history, is presented in my book *Nature: An Economic History* (Princeton, NJ: Princeton University Press, 2004). An excellent summary with new data on bite forces of large predators can be found in S. Wroe, D. R. Huber, M. Lowry, C. McHenry, K. Moreno, P. Clausen, T. L. Ferrara, E. Cunningham, M. N. Dean, and A. P. Summers, "Three-dimensional computer analysis of white shark jaw mechanics: How hard can a great white bite?" *Journal of Zoology London* 276: 336–342 (2008). The body masses of fossil and living predatory mammals are estimated in B. Sorkin, "A biomechanical constraint on body mass in terrestrial mammalian predators," *Lethaia* 41: 333–347 (2008).

Locomotion in large living amoebae and their probable Ediacaran ancestors is documented by M. V. Matz, T. M. Frank, N. J. Marshall, E. A. Widder, and S. Johnsen, "Giant deep-sea protist produces bilaterian-like traces," *Current Biology* 18: 1849–1854 (2008). For the evolution of marine burrowers see C. W. Thayer, "Sediment-mediated biological disturbance and the evolution of marine benthos," pp. 479–625 in M. J. S. Tevesz and P. L. McCall (eds.), *Biotic Interactions in*

Recent and Fossil Benthic Communities (New York: Plenum, 1983); M. L. Droser, S. Jensen, and J. G. Gehling, "Trace fossils and substrates of the terminal Proterozoic-Cambrian transition," *Proceedings of the National Academy of Sciences of the United States of America* 99: 12572–12576 (2002); and M. Kennedy, M. Droser, L. M. Mayer, D. Pevear, and D. Mrofka, "Late Precambrian oxygenation; inception of the clay mineral factory," *Science* 311: 1446–1449 (2006). Although the link between locomotion and nutrient exchange among ecosystems was not discussed by Polis and colleagues, the importance of such subsidies is recognized and well documented by them: G. A. Polis, W. B. Anderson, and R. D. Holt, "Toward an integration of landscape and food web ecologies: The dynamics of spatially subsidized food webs," *Annual Reviews of Ecology and Systematics* 28: 289–316 (1997).

Chapter 13: History and the Human Future

Important papers documenting increasing productivity under human agriculture include P. M. Vitousek, H. A. Mooney, J. Lubchenco, and J. Melillo, "Human domination of Earth's ecosystems," *Science* 277: 494–499 (1997); and R. F. Denison, E. T. Kiers, and S. A. West, "Darwinian agriculture: When can humans find solutions beyond the reach of natural selection?" *Quarterly Review of Biology* 78: 145–168 (2003). For a thoughtful account of human population increase, see Joel E. Cohen's *How Many People Can the Earth Support?* (New York: W. W. Norton, 1995). For an excellent source on human effects on soil, including the importance and consequences of irrigation, see D. Hillel, *Out of the Earth: Civilization and the Life of the Soil* (Berkeley: University of California Press, 1991). The history of the increasing effects of fire is well covered by D. M. J. S. Bowman, J. K. Balch, P. Artaxo, W. J. Bond, J. M. Carlson, M. A. Cochran, C. M. D'Antonio, R. S. DeFries, J. C. Doyle, S. P. Harrison, F. H. Johnston, J. E. Keeley, M. A. Krawchuk, C. A.

Kull, J. B. Marston, M. A. Moritz, I. C. Prentice, C. I. Roos, A. C. Scott, T. W. Swetnam, G. R. van der Werf, and T. W. Pyne, "Fires in the Earth system," *Science* 324: 381–384 (2009).

The case that resource depletion is not a foreseeable problem in the human economy is best put by Julian L. Simon, *The Economics of Population Growth* (Princeton, NJ: Princeton University Press, 1977). For the opposite view see Herman E. Daly, *Beyond Growth: The Economics of Sustainable Development* (Boston: Beacon Press, 1996).

The idea that most of the known biochemical diversity had arisen by the time true plants and animals had evolved is discussed by G. J. Vermeij, "Biological versatility and Earth history," *Proceedings of the National Academy of Science of the United States of America* 70: 1936–1938 (1973); L. Margulis, *Symbiosis in Cell Evolution: Life and Its Environment on the Early Earth* (San Francisco: W. H. Freeman, 1981); and D. T. F. Dryden, A. R. Thomson, and J. H. White, "How much of protein sequence space has been explored by life on Earth?" *Journal of the Royal Society Interface* 5: 953–956 (2008).

For a more detailed discussion of the human economy and its possible future see G. J. Vermeij, "Comparative economics, evolution and the modern economy," *Journal of Bioeconomics* 11: 105–134 (2009).

NOTES

Chapter 1

1. S. J. Gould, *Wonderful Life: The Burgess Shale and the Nature of History* (New York: W. W. Norton, 1989): 25.
2. D. S. Wilson, *Darwin's Cathedral: Evolution, Religion, and the Nature of Society* (Chicago: University of Chicago Press, 2002): 288.

Chapter 5

1. M. Rothschild, *Bionomics: The Inevitability of Capitalism* (New York: Henry Holt, 1999): 107.

Chapter 6

1. S. A. Kauffman, *Investigations* (Oxford: Oxford University Press, 2000): 35.
2. D. C. Dennett, *Darwin's Dangerous Idea: Evolution and the Meaning of Life* (New York: Simon and Schuster, 1995): 203.
3. R. Warren, *The Purpose Driven Life: What on Earth Am I Here For?* (Grand Rapids, MI: Zondervan, 2002): 25.

Chapter 12

1. S. J. Gould, *The Structure of Evolutionary Theory* (Cambridge, MA: Belknap Press of Harvard University, 2002): 1333.
2. C. de Duve, *Singularities: Landmarks on the Pathways of Life* (Cambridge, UK: Cambridge University Press, 2005): 235.
3. S. B. Carroll, *Endless Forms Most Beautiful: The New Science of Evo Devo and the Making of the Animal Kingdom* (New York: W. W. Norton, 2005): 288.

Chapter 13

1. G. D. Snooks, *The Laws of History* (London: Routledge, 1998): 107.
2. G. M. Hodgson, *Economics and Utopia: Why the Learning Economy Is Not the End of History* (London: Routledge, 1999): 108.

INDEX